FIELD GUIDE

TO THE

NATURAL

WORLD OF

WASHINGTON, D.C.

FIELD GUIDE
TO THE
NATURAL
WORLD OF
WASHINGTON, D.C.

Howard Youth

Illustrated by **MARK A. KLINGLER**

Photographs by **ROBERT E. MUMFORD, JR.**

Maps by **GEMMA RADKO**

Foreword by **KIRK JOHNSON**

JOHNS HOPKINS UNIVERSITY PRESS

BALTIMORE

© 2014 Johns Hopkins University Press
All rights reserved. Published 2014
Printed in China on acid-free paper
9 8 7 6 5 4 3 2 1

Johns Hopkins University Press
2715 North Charles Street
Baltimore, Maryland 21218-4363
www.press.jhu.edu

Library of Congress Cataloging-in-Publication Data

Youth, Howard.
Field guide to the natural world of Washington, D.C. / Howard Youth ;
illustrated by Mark A. Klingler ; photographs by Robert E. Mumford, Jr. ;
maps by Gemma Radko ; with a foreword by Kirk Johnson
    pages cm
  Includes bibliographical references and index.
  ISBN-13: 978-1-4214-1203-0 (hardcover : acid-free paper)
  ISBN-10: 1-4214-1203-9 (hardcover : acid-free paper)
  ISBN-13: 978-1-4214-1204-7 (paperback : acid-free paper)
  ISBN-10: 1-4214-1204-7 (paperback : acid-free paper)
  [etc.]
  1. Natural history—Washington (D.C.)—Guidebooks. 2. Parks—Washington
(D.C.)—Guidebooks. 3. Natural areas—Washington (D.C.)—Guidebooks.
4. Washington (D.C.)—Guidebooks. I. Klingler, Mark A. II. Mumford, Robert E.,
1935–. III. Title. IV. Title: Field guide to the natural world of Washington, D.C.
  QH105.W18Y68 2014
  508.753 — dc23   2013015239

A catalog record for this book is available from the British Library.

All watercolor plates © 2014 Mark A. Klingler
All photographs © 2012 Robert E. Mumford, Jr., except pages 215 and 253,
  courtesy Wikimedia
All maps © 2014 Gemma Radko

*Special discounts are available for bulk purchases of this book. For more
information, please contact Special Sales at 410-516-6936 or specialsales@press
.jhu.edu.*

Johns Hopkins University Press uses environmentally friendly book materials,
including recycled text paper that is composed of at least 30
percent post-consumer waste, whenever possible.

Book design by Kimberly Glyder

CONTENTS

FOREWORD

AFTER 30 YEARS OF COMING TO WASHINGTON, D.C., to visit the Smithsonian's National Museum of Natural History (NMNH), last fall I became its director. With more than 7 million visitors and 127 million objects, NMNH is the largest and most visited nature and science museum in the world and, certainly, a required stop for any naturalist living in or traveling through the nation's capital. Yet my 30 years of visiting Washington, D.C., revealed other places to see nature. I had brushing encounters with the Potomac River, Rock Creek Park, and the National Arboretum and—even while challenged on occasion by the swampy mugginess of the city—remained impressed with its profusion of lush green pockets. As I walked to the museum, I would often spot birds, bushes, or trees that were novel to my Denver-trained eyes. Now that I live here, and as the nation's chief natural historian, I find myself in a need-to-know situation with respect to local nature.

Within weeks of arriving in Washington, I heard from my friend Mark Klingler, who was just completing the illustrations for a book about the natural history of my new home. Mark had teamed with author Howard Youth and photographer Robert Mumford, Jr., to create a neat and portable volume about the nature that exists in my new city. Using a combination of paintings, photographs, and prose, they tell tales of the city's parks and paths and do a great job of exploring and exposing the biodiversity that permeates the city. Located at the confluence of the Potomac and Anacostia Rivers at the head of Chesapeake Bay and just downstream from the Fall Line that marks the edge of the Appalachian Piedmont, Washington is a natural crossroad for migratory birds and a great place to experience the forests of the East Coast. The book describes the best locales to escape from sometimes frantic city life, places that provide opportunities for close encounters with nature.

Among the places highlighted in the book are D.C.'s wetlands. People often say that Washington, D.C., used to be a swamp, as if that is a derogatory statement. Yet, as this book shows, Washington D.C.'s swamps are some of the best locations to see amazing plants and animals. This book is a fantastic portal to them and to the rest of the natural world that surrounds and intertwines with our nation's capital.

Kirk Johnson
Sant Director, Smithsonian National Museum of Natural History

ACKNOWLEDGMENTS

THIS BOOK IS A CELEBRATION OF NATURE in one of the country's prettiest and most important cities. But here at the front, I'd like to celebrate the many people who shared their knowledge and skills with me. Without them, I could not have written this book.

With artistic mastery, Mark A. Klingler breathed life into a staggering array of life-forms so you could see the wildlife I describe in the text. Rare is the artist who can capture the sophistication of a raccoon and the simplicity of an earthworm on an oak leaf. Thank you, Mark, for sharing your talents with us.

Two longtime friends and fellow nature enthusiasts made great contributions to this book.

Robert E. Mumford, Jr., took all of the book's photos. Readers of this book will benefit from Bob's dogged wanderings and high photographic standards. He traveled around the city to make sure all plant and rock photographs were not only up to my standards but also his. Bob has created a stunning collection of photos of the parks and wildlife of D.C., many of which appear in his own book, *Spring Comes to Washington*. This field guide would not be the same without his contributions.

Robert E. Mumford, Jr., thanks the following people for helping to locate trees and shrubs to photograph: Gail Mackiernan, Mary Kay Smith, Barbara Knapp and Angela Butler of Butler Orchards.

Gemma Radko produced all of this book's maps. She also put me in touch with important contacts, provided general information, and proofed my early drafts—all this done after her daily work for the American Bird Conservancy. The maps provide visual orientation no words could capture and should make park visits much easier for readers. Thank you, Gemma, for all your work. You deserve a birding break!

Along the way, many people received my calls and e-mails and shared their time and insights, reviewing sections, answering questions, or providing further resources. Thanks go to Melissa Cronyn, chief of National Park Service Publications, and her colleagues, for putting us onto the fine online collection of National Park Service maps.

A big thank-you to Callan Bentley at Northern Virginia Community College for demystifying the city's geology and making it come to life and for reviewing my draft material and making it better and more accurate. My friend David Grimley gave me early encouragement as I first embarked on reading up on geology. Thanks also to

William C. Burton, U.S. Geological Survey, for his guidance and wise words regarding the city's geology.

I thought I'd seen it all, but Jason Berry opened my eyes to new birding strategies when he spoke with me about his big year biking and hiking for birds in the District. Jason also followed up with important information on locations and organizations, which I have shared here with you. Fort Dupont Park is included as a separate section thanks in good part to Jason's exploration and reporting on this hidden gem.

Thanks also go to fellow birders Michael Bowen, Greg Gough, Rob Hilton, Paul Pisano, and Bill Young for their input regarding D.C. birding hot spots and the status of certain species; to rangers Alexis Gelb and Kate Bucco at Anacostia Park and Kenilworth Aquatic Gardens; Rock Creek Park's Ken Ferebee for his insights not only on Rock Creek Park but also on other nearby parks; Nancy Luria and Susan Greeley at the National Arboretum; Damien P. Ossi, Dan Rauch, and Lindsay Rohrbaugh at the District Department of the Environment's Division of Fisheries & Wildlife; Wayne Hildebrand at the North American Amphibian Monitoring Program; Jay Kilian of the Maryland Department of Natural Resources; and Matt Cohen and Rob Gibbs (information about local fungi). To the others who helped along the way, and helped me find these valued resources, thanks.

Amy Chang and Bob Young, who are dear neighbors, friends, and kindred nature spirits, provided not only their encouragement but also notes on their special exploration of Glover-Archbold Park. They also guided me to helpful information on hiker/biker trails.

Many years ago, in the 1970s, lifelong family friend Marilyn Blakely gave me my first pair of binoculars. When she heard I was working on this book, out of the blue she mailed me a copy of Louis J. Halle's wonderful *Spring in Washington*, a snapshot of nature in the capital during the 1940s. Throughout the research and writing phases, this book reminded me that while the city has lost some wild places and creatures, many of its natural assets remain.

Thank you to Susan Lumpkin, friend, former boss, and writer and researcher extraordinaire for recommending me for this job; for teaching me how to be a better editor, writer, and researcher; and for reviewing the mammals section of this book.

The Johns Hopkins University Press team made this book happen. Thanks to Vincent J. Burke, Andre Barnett, Jennifer Malat, Sara Cleary, Catherine Goldstead, and Laura Ewen.

In the course of researching this book, my admiration for fellow authors grew tremendously. Book writing is a complex journey, like a river ride down rapids, where you get caught in eddies and work your way back on course. For their detailed and lovingly researched treatises, I would like to thank the late Claudia Wilds for her meticulously researched *Finding Birds in the National Capital Area*, Mark Garland for his book *Watching Nature: A Mid-Atlantic Natural History*, and John Means for the excellent *Roadside Geology of Maryland, Delaware, and Washington, D.C.*

Words won't adequately express my gratitude to family and friends who supported and encouraged me throughout the process. To the now widely dispersed Writer's Group of Quito—Amanda Fernandez, Francesca Contiguglia, Wes Carrington, and Marta Youth: I greatly enjoyed the wine, encouragement, commiseration, and advice.

I've felt a tugging curiosity about the natural world since at least the age of three. Though not nature lovers themselves, my parents saw that animals and plants inspired me, so they always supported the growth of my interests, from when I was a boy tending a growing collection of lizards, newts, and hermit crabs and digging holes in the yard to plant wildlife-attracting plants to when I was a young man obsessed with birding local parks and beyond. Thank you, Dad and Mom, for all your love and sacrifice, and for recognizing and supporting my interest in nature so I can now share it with others. And thank you for your trust, handing me the keys on my sixteenth birthday so I could embark on my first solo birding adventure. As a parent now, I know how hard that must have been for you. Thanks also to my sisters for putting up with my wildlife pets and interests over the years. To Youths Gail, Zachary, Alessandra, and Thomas, thanks for their special exploratory foray to Roosevelt Island with me.

Deepest gratitude goes to my wonderful wife, Marta, who was a tremendous support through the whole process, and to my kids, who always thought it was great that dad was writing this book. And finally, a pat-pat to our Andean crumbhound Annie, who faithfully kept me company while I pounded out the text for this book. Here it is!

H.Y.

To my loving parents, my precious wife, and my wonderful children.

H.Y.

❋

I would like to thank my wife, Cathy, for her constant support as I worked on the book, and my daughter, Olivia, for the impromptu drawing sessions. To my Mom and Dad, thank you for encouraging me in drawing and sculpting nature in the early years. I trust you both would be pleased. Also, my gratitude to all of my family and friends for their extensive knowledge and for finding critters for me to photograph and draw. Thank you, Howard, for photo references of locations in the parks and for the opportunity to share images of so many of D.C.'s natural wonders with the public. Vincent J. Burke, thank you for your insights and for keeping me to deadlines so we could create this learning tool.

M.A.K.

the natural, and not-so-natural, history of washington, D.C.

TWO GEOLOGICAL PROVINCES AND TWO different rivers help make our striking capital city an urban ark for varied flora and fauna. Down Washington, D.C.'s west side runs the Fall Line, where rolling Piedmont cedes to a gradually leveling Coastal Plain. An easy place to experience this transition is at Roosevelt Island, which sits in the Potomac between Georgetown and Rosslyn, Virginia. The north side of the island fronts rocky shore, while tidal wetland fringes the island's south side. About 3 miles southeast, the Potomac River merges in a watery Y with one of its main tributaries, the Anacostia River. These convergences—the Fall Line and the two rivers merging—

Cedar waxwings flock to ripe berries. Though these showy birds visit or nest in many of the city's parks, they often go unnoticed.

plus the meeting of northern and southern influences make the city part of a floral and faunal melting pot. The location's natural assets also attracted George Washington's attention. He picked this spot for the capital city that would later bear his name.

Other eastern cities began and grew along the Fall Line where Piedmont meets Coastal Plain. There, commerce was most vibrant because of a natural win-win situation: easy navigation from ocean to tidal Coastal Plain river ports and an abundance of moving water to power mills at rocky falls where rolling Piedmont ceded to Coastal Plain. Baltimore, Philadelphia, Wilmington, Richmond, and Raleigh are all Fall Line cities.

The Potomac has its origins in the Appalachians and defines much of Maryland's squiggly western boundary with northern Virginia and West Virginia. After the river flows south of Washington, D.C., it widens and takes on a leisurely Coastal Plain pace before reaching the country's largest estuary, the Chesapeake Bay, about 100 miles southeast of the city. The Potomac is the Chesapeake's second-largest freshwater source, after the Susquehanna.

The Potomac is also the city's principal source of freshwater, a rebounding sport fishery, and a focal point for recreation in and around the nation's capital. The river seems to keep its own time, flowing past the hyperactive capital city perched at one of its wide bends.

The rocks and soil we shift around in our gardens tell a story of upheaval and wearing that help dictate the city's water flows and soils and thus its wildlife. Before the dinosaurs, between 320 and 250 million years ago, North American and African landmasses collided during the formation of the supercontinent Pangaea. (Pangaea began to split apart 50 million years later.) North America's eastern edge was gradually pushed up to form a dramatic mountain range resembling the Alps or Rocky Mountains. Intense heat and pressure kneaded rocks into a wavy jumble and transformed sedimentary and igneous rock into metamorphic rock.

Even as it was formed, the mighty mountain range and its foothills, or Piedmont, began to weather. The uneven erosion of varied rocks formed the hilly landscape and rugged river valleys we now enjoy in the city's parks. Yet as we hike along the Potomac and through woods along its tributaries, how often do we ponder the tiny sediments washing past us, which may hail from the mountains and are on their way to build the Coastal Plain?

Before settlement, expansive, towering forest cloaked much of what is now the nation's capital. Below the Fall Line, open marshy expanses hugged the Anacostia and lower Potomac shores. The rivers abounded in herring, shad, salmon, and sturgeon. Waterfowl nested and transited these watery highways on northbound and southbound journeys in far greater numbers than today. Wolves and pumas hunted deer and perhaps eastern elk in the area.

Native Americans were drawn to the area's natural riches and its location near two main trade routes. Archaeological evidence dates human presence to as far back as before 1500 BC. By the early 1400s, the Conoy or Kanawha, of Algonquian stock, settled along the Potomac. They hunted, fished, and grew corn in the rich river soils. River villages dotted the river shores by the early 1600s. Nacotchtank, the main village of the Nacotchtank, or Anacostan, tribe sat near the Anacostia River's east bank, near current-day Anacostia Park.

The 1600s brought great change to human populations. Tribal conflicts became more frequent and intense, and English settlement moved north from the Jamestown colony, bringing with it more conflict, land-use changes, introduced disease, and increased trade that included beaver pelts and firearms. European settlements spread and Native American villages vanished by the end of the 1600s.

By the early 1700s, tobacco drove the economy and transformed the landscape. Growers cleared most of the forest to make way for this important cash crop. Corn and wheat and cattle and pigs were other important commodities.

Following the Revolutionary War's end in 1783 and the Constitution's creation in 1787, Washington was tasked with selecting the site for the new nation's capital city. He chose an area where the Anacostia flowed into the Potomac, just east of the town of Georgetown, Maryland (now part of D.C.), and northeast of Alexandria, Virginia, and his home at Mount Vernon.

The distinctive bark of the American sycamore.

In 1791, Pierre-Charles L'Enfant, a French-born architect and civil engineer hired by Washington, presented his plans for the newly declared capital. L'Enfant's city would have European elegance. Grand tree-lined avenues would provide sweeping views of a capitol and the president's home. Most streets, though, would follow an orderly grid pattern.

At the time, George Washington, L'Enfant, and others imagined the capital city would have a vibrant port. The Washington City Canal was dug to link the rivers and to run boats through the city. However, floods, silt, and tidal fluctuations would, over the years, dictate otherwise. In the 1870s, the canal was closed and filled.

Planners have doggedly tried to keep L'Enfant's plan alive. In 1901 and 1902, the Congress-appointed McMillan Commission developed a plan to rejuvenate and expand the city's parks and monuments. It took decades to complete the proposed projects, which resulted in many of the parks described in this book. Today, a commitment to L'Enfant's plan dictates management of the city's downtown open space and building patterns, including height restrictions that maintain the city's unique skyline.

According to the Trust for Public Land, parkland covers 19.4 percent of Washington, D.C., making it the country's second-most park-rich large city. This figure includes 7,617 acres in National Park Service and city properties but does not include other large areas

White-tailed deer live in most of the city's wooded parks. They are often seen during the morning and near dusk.

such as the U.S. Department of Agriculture's National Arboretum and the Smithsonian's National Zoo.

In 2002, staff from the nonprofit tree-conservation group Casey Trees and the National Park Service, along with hundreds of volunteers, surveyed D.C.'s urban forest. They estimate that about 1.9 million trees grow within the city boundaries, including about 106,000 street trees. The most prevalent street tree species were red, Norway, and sugar maples and pin oak. Within the forest, American beech, red maple, boxelder, and tuliptree were the most common species. According to 2006 U.S. Forest Service data, 35 percent of the city's area is covered by trees. The city's Department of Transportation runs an urban forestry program under which thousands of new street trees are planted each year.

Over the years, some of the city's wildlife has disappeared, including the northern bobwhite quail and eastern meadowlarks that used to nest in brushy and grassy areas. Other species were once far more common than they are today, including wood thrush and ovenbird. Furthermore, white-tailed deer and Canada goose are far more common than in the past. Other species are recent arrivals. The coyote, for example, finally found its way into Rock Creek Park after decades of eastward range expansion. The introduced northern snakehead, a fish native to Asia, now lurks in the Potomac, preying on other fishes. Even the crayfish and lady beetle communities are changing.

Perhaps the greatest change has come to the city's flora. The plant community in many parks and gardens would surprise past naturalists if they were able to visit today. A wide range of nonnative invasive plants flourish, while native plants feel the pinch with the city's hungry and growing deer herd.

Through it all, Washington, D.C., remains a fascinating and challenging place for the naturalist. Beautiful trees and flowers are usually in view, and even during winter, you can escape to large parks and feel a wildness rarely experienced in other capital cities.

Recent surveys highlight the city's biodiversity. In 2007, a 24-hour BioBlitz survey of Rock Creek Park recorded 661 plant and animal species. A similar survey of Kenilworth Recreation Park and Kenilworth Aquatic Gardens in 1996 found about 975.

A temperate meeting place for northern and southern flora and fauna, the Washington, D.C., area feels and appears almost tropical in summer and can be quite mild in winter, depending on the year. In some years, though, it receives more than its share of ice or snowstorms. Rarely does the ground remain frozen for long periods.

Rich soils and an average rainfall of about 40 inches provide the palette on which this garden city was created. Trees grow tall in Washington, adding to the established, solid feel of this capital city. Flowers, bushes, and vines flourish. From the start, planners incorporated these living embellishments at the U.S. Capitol, the White House, in the memorial parks and the National Mall, and in the larger, more natural parks.

Washington is a city of politicians, lawyers, lobbyists, agencies, embassies, and media and is often viewed as a high-stress place. Millions visit each year for work or for pleasure. But Washington is also a city of cyclists, joggers, hikers, birders, anglers, rollerbladers, and people with many other outdoor hobbies. For those who love nature, the city's green welcome mat is obvious. The parks host a changing tableau of blooms and a changing roster of migratory birds. They are great places to see fireflies' nocturnal light show or to hear the annual cicadas' sizzling summer chorus. I hope this book will help residents and visitors look more closely at the wildlife that shares with us the nation's capital—from pines to pillbugs to pileated woodpeckers.

2

visiting
D.C.
parks

WASHINGTON, D.C., IS A CAPITAL with many easily accessible natural areas—towering forests, river shores, gurgling streams, and sprawling gardens. You might think the city was designed with naturalists in mind. Perhaps grandeur was more the goal, but in either case, the result is exciting for anyone interested in the natural world. In this book, I describe a baker's dozen of the most promising places to view Washington, D.C., wildlife, noting other nearby areas as well.

If you think all of these areas are well known to Washingtonians, think again. Worldly, wise, and always short on time, city residents and daily commuters from Maryland or Virginia typically miss all but what's in front of them on the sidewalk or parkway. Sometimes they miss that, too. Then, when it's finally time to relax, where do they go? As Mike Tidwell wrote in an article on the merits of fishing in the city, which appeared in the *Washington Post* in 1998: "Escaping to nature for most people means putting as many miles between themselves and the Beltway as quickly as possible."

I have visited many city parks in many parts of the world, and the more I travel, the more I appreciate what I left at home. I grew up in Silver Spring, Maryland, just north of the city. As I've researched these parks, reading their histories and the reflections of past and present visitors, I am thankful for all the green space we enjoy today. These places were made possible because of the vision of presidents, first ladies, congressional representatives, and such champions as banker Charles Carroll Glover and others. (For a discussion of Glover's legacy, see page 59.) They left us an incredible legacy in inspirational places to explore.

Inside parks like Rock Creek Park, the C&O Canal National Historical Park, or Fort Dupont Park, you feel as though you've escaped to the country. Recreation and relaxation are among the primary purposes of urban green space. But the convenience of the city and its parks can easily lull us into thinking we don't need to prepare for these outings as we would a longer trip. A little preparation goes a long way toward making your D.C. park visits as enjoyable as possible.

Pack patience with you on any excursion. Washington, D.C., traffic ranks among the most challenging in the country. The city has many one-way streets and traffic circles. The parkways are curvy and scenic but vexing for newcomers, as lanes suddenly appear and

Benches near the Vietnam Veterans Memorial and Constitution Gardens await visitors.
The city is full of parks where you can relax, reflect, and enjoy nature.

disappear, and signs for important turnoffs pop up at the last second. Read parking signs carefully to avoid getting a ticket.

The iconic U.S. Capitol is more than just a meeting place for Congress. Like a domed hubcap, it provides the central point from which invisible perpendicular lines radiate in four directions, dividing the city into four different-sized quadrants: Northwest, Northeast, Southwest, and Southeast. Three of these lines are traced by North Capitol, East Capitol, and South Capitol Streets; the other runs west, lengthwise down the National Mall's middle.

Here's a tip I learned the hard way as a teen driving in the city: You may want to pay attention to those compass orientations at the end of street names. In the text, I try to include N.W., N.E., S.E., and S.W. after most city streets. In Washington, numbered streets usually run north–south, with numbers going up the farther you walk or drive east or west from the U.S. Capitol. So, 5th Street, N.W., is an entirely different street from 5th Street, N.E. Yet if you drive from 5th Street, N.E., south to 5th Street, S.E., you are on the very same road. It just changes quadrants when you cross East Capitol Street. This all makes sense when you look at a map and remember the U.S. Capitol is the city's central starting point.

To avoid car travel, when feasible, walk or use the Metrobus or Metrorail systems. You can work out itineraries online at Metro OpensDoors.com using the trip planner option. If you are walking, travel during daylight and in well-traveled, open areas.

Use common sense as you explore the city's parks. Travel with others, especially in unfamiliar areas for security and to help you safely navigate to the right road or trail. Take your cellphone. Parkland described in this book covers thousands of acres. Even though most trails are well marked, getting lost is not impossible. Park personnel ask visitors to stay on designated trails, but many unplanned and unmarked side trails crisscross the city's parks.

Many locals gripe that Washington, D.C., has just two seasons—"tropical" summer heat and humidity and either icy, snowy, or muddy winter. Washington, D.C., has four distinct seasons. They are just not evenly distributed. It's one of those places where people obsess over meteorology. Checking a weather forecast is always a good idea before you head out so you know how to dress, whether you should bring a hat or raincoat, and so on.

Pack sunscreen, ample water or other drinks, and snacks. Summer heat can sneak up on you, and many of the parks in this book do not have refreshment stands or at least ones easy to reach quickly. Most have restrooms and visitor facilities, but these may not always be close by or open, depending on where you wind up exploring. If locating restrooms is a priority for you, you may want to contact the park headquarters for sure-fire locations. There is contact information in each of this book's park accounts.

Consider showering and changing clothes after you hike on or off narrow trails in fields, brush, and forest. Lyme disease continues to rise in the area, and deer ticks are notoriously tiny. I always make it a policy to shower when I return from a morning or day in the field, whether in D.C. or outside it. Of course, ticks are not active in freezing temperatures.

Ticks aside, another invertebrate that captures attention in and around the city is a new arrival, the Asian tiger mosquito. From late spring to early fall, these insects breed in small amounts of stagnant water and, unlike other local mosquito species, they are active throughout the day. I rarely need repellent when visiting the city's parks, but if you are sensitive to insect bites, having some on hand is a good idea. For those with sting allergies, be aware that yellow jackets top the city's list of close-up creatures during summer and early fall. They frequent outdoor food stands and trash cans.

Any chance you will see a poisonous snake in the nation's capital? Probably not. Most snakes there are nonvenomous, such as the northern brown snake, eastern garter snake, black rat snake, or northern water snake. Common in many parks, the northern water snake is often misidentified as a venomous copperhead or a cottonmouth. Cottonmouths, also called water moccasins, do not occur in the region; southeastern Virginia is the closest they get. Around the Beltway, the northern copperhead is the only local venomous snake but lives primarily in Maryland and Virginia parkland. The city's wildlife department lists northern copperheads as "critically imperiled." There have been no confirmed copperhead sightings in Rock Creek Park since the 1970s. However, there are periodic reports along the rivers in areas such as the C&O Canal, so it might be wise to watch where you step or reach.

What should you bring on an outing to document wildlife sightings? Nothing helps confirm a wildlife sighting faster than a good photograph. Photos also allow you to "take home" sightings so that, later on, you can sit down with some books and puzzle over tricky identifications. Binoculars bring you close views of birds, amphibians and reptiles, butterflies, dragonflies, and even distant leaves and wildflowers. Have you ever watched a chipmunk or squirrel through binoculars? You see so many details you might otherwise miss. Keep a pencil or pen and notebook or sketchbook handy. Then you can write down details you might later forget, keep lists, or jot down location or weather details. I always keep detailed notes and find myself referring to them year after year to compare seasonal events, such as when I see the first eastern kingbird or the first flowering dogwood blooms of spring.

Washington, D.C., is a great place to explore, whether you have just a weekend or a lifetime. If you find yourself with a weekend or a weekday morning or afternoon and feel like escaping the pavement and crowds, you don't have to hurry off to a distant countryside. Just find one of these parks and escape the city inside the city.

Cherry blossoms are not the only game in town: flowering magnolias and redbuds on the National Mall in spring.

CHAPTER 3

the
parks

NORTHWEST

Rock Creek Park

LOCATION

From the boundary line with Maryland, the National Park Service–administered Rock Creek Park and Rock Creek and Potomac Parkway snake south down the city's northwest quadrant, ending about 9 miles south at the Potomac River.

Nature Center (Wed. to Sun., 9 a.m. to 5 p.m.; closed holidays)
5200 Glover Road, N.W.
Washington, D.C. 20015

TELEPHONE

(202) 895-6070

WEBSITE

http://www.nps.gov/rocr/index.htm

SIZE

The city's largest park, at 1,750 acres. (It is more than twice the size of New York City's Central Park.) Rock Creek Park is 9.3 miles long and up to a mile wide, but much more narrow in many areas.

HABITATS

mature and second-growth hardwood ridge and riparian forests, grassy fields, creek, shrubby forest edges

Natural History   Over the millennia, Rock Creek etched a snaking, ridge-lined gully through resistant metamorphic rock in what is now the Northwest quadrant of the city. Running down this hilly park's east side is 16th Street, which roughly marks the Fall Line, where the flatter Coastal Plain begins. The bottom of the park, where Rock Creek spills into the Potomac across from Roosevelt Island, marks the area on the river where Piedmont yields to Coastal Plain.

Rock Creek starts more than 30 miles to the north, with its source in Laytonsville, Maryland. The creek, in turn, feeds into the wide Potomac at Georgetown. The Potomac River is the second-largest freshwater source feeding into the Chesapeake Bay, which is the nation's largest estuary.

Most of the park is cloaked in forest. Beech–white oak forest covers the most area. Within this habitat, American beech, eastern white oak, and tuliptree dominate the canopy, joined here and there by several other oak species. Small trees and shrubs grow beneath the canopy, including flowering dogwood, American holly, and mapleleaf viburnum. Mayapple, jack-in-the-pulpit, and the nonnative garlic mustard are among the wildflowers, while nonnative vines such as oriental bittersweet and mile-a-minute climb the trees, especially in sunny clearings.

Tuliptrees predominate in fairly moist areas where past clearing and soil disturbance were severe. At these sites, boxelder and spicebush may also flourish. In places, wineberry and multiflora rose shrubs and porcelainberry vine dominate.

Well-drained, undisturbed ridgetops and slopes feature drier forests dominated by chestnut oak and black gum. Red and black oak may be found in this forest, along with shrubs such as blueberry, huckleberry, mountain laurel, and native azalea.

Made of local stone, Boulder Bridge blends with Rock Creek's picturesque shores.

Narrow corridors of floodplain forest grow at the edge of Rock Creek and its associated streams. American sycamore is an indicator of this habitat, and boxelder, red maple, tuliptree, green and white ash, and river birch are other trees found there. Jewelweed, jack-in-the-pulpit, and other native wildflowers grow here, as well as invasive nonnative plants such as fig buttercup (or lesser celandine), garlic mustard, and Japanese stilt grass.

HUMAN HISTORY In 1608, Captain John Smith explored the Potomac River. In his journal, he mentioned a creek spilling into the river at the site where the Thompson Boat Center now sits. He was the first European to describe Rock Creek.

In 1862, 28 years before the park's establishment, large tracts of forest were cleared from what is now the northern part of the park. This forest was yet another casualty of the Civil War. Union soldiers and local laborers cut the trees to make way for a supply and access road known today as Military Road, to provide sight lines between new forts built to protect the city from Confederate attack and to provide clear lines of fire for the cannons of Fort DeRussy. All told, 68 forts ringed the city, including Fort DeRussy, the mounded remnants of which lie in the park at its highest point. With the forest cleared between forts, Union forces could see flags flying, invading Confederate troops kicking up dust, or smoke from cannon fire.

The park administers other fort sites in the city's north sector, including the scant remains of Fort Reno, highest point in the city, Fort Bayard, and Fort Stevens—the city's northernmost fort and site of the only Confederate attack on the city in July 1864. Abraham Lincoln watched some of the battle from that fort's parapet.

In 1866, the year after Robert E. Lee's surrender at Appomattox and Lincoln's assassination, the quest to establish a large city park in the capital began. By that time, a housing boom was under way, but much of the current park area remained forested because of its steep topography. The park was created in 1890, following 24 years of lobbying, fund procurement, and land acquisition. President Benjamin Harrison signed off on the bill that established Rock Creek Park, 32 years after New York's Central Park first opened. Like that urban gem, Rock Creek Park was a much-needed refuge from the muddy, grimy, smoky streets of late-nineteenth-century Washington. However, unlike Central Park's largely human-created landscape, Rock Creek Park remained mostly untouched, with a maturing forest and a rolling stream, a natural masterpiece later embellished with bridges and other structures.

Rock Creek Park is one of the oldest national parks and one of the largest and most wild city parks in the country. In a city of parks, it is the park of parks. One park employee calls it "the Capital City's Forest." Within its borders grow some of the oldest deciduous forest left in the region.

Over the years, Rock Creek Park has been a wild playground for many nature-loving Washingtonians, including presidents. John Quincy Adams enjoyed listening to bird song there. Teddy Roosevelt scrambled up and down the wooded slopes, challenging friends to follow. He also rode his horse in the park.

In 1918, Frederick Law Olmsted, Jr.—a leading landscape architect and son of the designer of New York City's Central and Prospect parks—oversaw a study of the park that fleshed out core management goals still carried out today in this urban wild space. According to the report, the park's "interesting, varied, natural scenery must be saved intact insofar as possible, must in some respects be restored or perfected by intelligent, appreciative landscape development, and must not be replaced by other and more or less foreign types of 'treatment.'"

The report called for buildings, bridges, and other structures that were "harmonious and subordinate parts" that would not detract from the wild setting. Many structures, including the Boulder Bridge along Beach Drive, were made of local stone that blends with the rocky setting. The primary goals were to protect the serenity and beauty of the park's rugged creek and its valley and to keep the park looking "undeveloped." Today, the National Park Service strives to balance these goals with Washingtonians' growing recreational needs.

Rock Creek Park runs from the Maryland Line south to the Potomac River, providing a corridor for the dispersal of native plants, fish, amphibians, reptiles, and mammals. It is the city's largest block of forest, and its most important migration stopover for Neotropical migrant birds, including warblers, orioles, flycatchers, and cuckoos. Throughout the park, watch for red-shouldered hawks, barred owls, and pileated woodpeckers. By the streams, you may see wood duck, belted kingfisher, or hear eastern screech-owl. From late spring into summer, breeding woodland songbirds include eastern wood-pewee, Acadian flycatcher, ovenbird, and scarlet tanager.

The park's reptiles and amphibians usually don't present themselves, but you may find one or two of these nonvenomous snakes: black rat snake, eastern garter snake, northern water snake, northern brown snake, and northern ringneck snake. Since the 1970s, there have been no confirmed sightings of the venomous northern copperhead, which was once present in the park. Five-lined skinks may be found around logs, rocks, and sunny patches near tree trunks. Chance encounters with American toads, eastern box turtles, and redback salamanders are possible in the woods. In the creek's wider spots and along its shoreline, eastern painted turtle and red-eared slider may turn up.

Expect to see gray squirrels, eastern chipmunks, and white-tailed deer. Other mammals that live here but are less likely to be seen include northern raccoon, opossum, red and perhaps gray fox, and the newly arrived coyote, which was first seen by park staff in 2004.

Concentrated within the long, thin park, the park's white-tailed deer overbrowse and overgraze the woods. Their intense quest for food greatly diminishes undergrowth—tree saplings, shrubs, and wildflowers. Disturbance to the soil and the deer's seed-laden hooves help spread Japanese stiltgrass and other invasive nonnative plants that overwhelm many native species. According to the National Park Service, "since 1991, data gathered from the park's vegetation monitoring program clearly show that nearly all tree and shrub seedlings are being browsed by deer before they have a chance to grow." The park's deer management plan aims to protect vanishing native plants while monitoring and reducing

By March, wood duck pairs become a familiar sight along Rock Creek.

the park's deer population. Despite these pressures, the park remains an important refuge for the city's flora and fauna.

Fishing is prohibited north of the Porter Street Bridge, but below it, anglers who frequent the park know where to find catfish, bass, carp, and other fish. At Peirce Mill, a fish ladder now allows migrating herring to bypass the manmade falls there and to continue upstream to spawn. This 2006 remediation removed a fish migration barrier that stood for more than a century.

Beach Drive at the Maryland State Line South to Peirce Mill
Beach Drive is the principal road winding from Maryland (in Rock Creek Stream Valley Park) down into Rock Creek Park. South of the Smithsonian's National Zoo, the road changes name to the Rock Creek and Potomac Parkway. The entire route passes through forest and by playing fields, and Rock Creek is in sight along much of its course. The northern part of the park has swampy woods thick with skunk cabbage by March. During migration, birders should stop at picnic areas and watch for mixed flocks of birds. During weekends (7 a.m. Sat. to 7 p.m. Sun.) and holidays, three road closures are in effect, so cyclists, walkers, joggers, and skaters can enjoy the park. (See pages 47 and 48.)

Nature Center to Peirce Mill To enter the park off Military Road, N.W., turn south on Glover Road, N.W. Once on Glover Road, N.W., you will see the old military field on the right, now maintained as meadow, supporting butterflies, field birds, and wildflowers.

The Nature Center has many programs focused on local wildlife. Check here for information on the park, other nearby National Park Service properties, and the varied program schedule. There is a

East West Highway

Grubb Rd

410

Meadowbrook
Riding Stable

Beach Dr

Westbrook La

Primrose Rd

East Beach Dr

W Beach Dr

North Portal Dr

South Portal Dr

ROCK CREEK PARK
(MD-MCPPC)

Leland St

Candy
Cane
City

Boundary
Bridge

Beach Dr

Parkside Dr

Rock Creek

17th St NW

Holly
Rd

East West Highway

Daniel Rd

Pinehurst
Pkwy

Western Ridge Trail

Wise Rd

Connecticut Avenue

Chestnut   St NW

Beach St NW

Riley Spring
Bridge

Aberfolye Pl NW

31st St NW

ROCK
CREEK
PARK

10

MARYLAND
DISTRICT OF COLUMBIA

Western Avenue

Tennyson St NW

Oregon Ave

Sherrill
Dr

CHEVY CHASE

Rittenhouse St NW

Bingham Drive

Rolling Meadow
Bridge

16th St NW

Chevy Chase
Circle

Park
Police
Stables

Valley Trail

PUBLIC
GOLF
COURSE

30th St NW

Newlands St NW

Milkhouse
Ford

Miller Cabin

Joyce Rd

17th St NW

Nebraska   Avenue NW

Fort
DeRussy

Military Rd NW

M Friendship
Heights Metro

Military Rd NW

LITTLE
FOREST
PARK

Nature Center and
Planetarium

29th St NW

Connecticut

Reno Rd NW

Information

Horse Center

Ave NW

Wisconsin Ave NW

Fessenden   St NW

Maintenance Yard

Rock Creek

Morrow Dr

16TH &
KENNEDY
AREA

Ballfields

River Road NW

Belt Rd NW

Broad Branch Rd NW

Grant Rd

Glover Rd

Western
Ridge
Trail

Rapids
Bridge

Tennis
Stadium
and Courts

FORT RENO
PARK

Chesapeake St NW

Parking

Box Office

Brandywine St NW

Ross Dr

Broad Branch

17

Carter Barron
Amphitheatre

Albemarle St NW

18

Boulder
Bridge

Beach Dr

Colorado Ave NW

Blagden Ave NW

17th St NW

29

Tenleytown
Metro

Tenley
Circle

M

29th St NW

Equitation
Field

Audubon Ter NW

Van Ness-UDC
Metro

SOAPSTONE
VALLEY
PARK

Pulpit
Rock

Van Ness St NW   M

planetarium here as well—the only one set up and run by the National Park Service.

Sometimes milkweed is planted in front of or behind the Nature Center, attracting monarchs, which lay their eggs there. Milkweed is the monarch's host plant, providing the main food for its caterpillars. This plant also attracts feeding great spangled fritillaries, silver-spotted skippers, and others. Also, watch for monarchs in the park's fields. From September to early October, migrating monarchs flap and sail over the treetops.

During migration, the Nature Center and its wooded parking lot attract migrating woodland birds. Two trails leave from the Nature Center, including a small loop trail that can be good for migrating thrushes. Watch the feeders and birdbath for an interesting mix of songbirds and perhaps woodpeckers.

The tall trees around the parking lot can provide good views of scarlet tanagers, various flycatchers, yellow-billed and black-billed cuckoos, and any of the eastern warblers. These birds may also be seen at the Ridge, the Maintenance Yard, and other spots.

Near the parking lot are the horse stables. Just outside the back of the stables, a bridle trail heads into the woods, leading on the right to a short path that spurs off to an opening known as the Maintenance Yard. The yard is on a rise overlooking the forest edge. In one spot, tall stacks of stone blocks from the U.S. Capitol sit amid the trees and bushes, likely sitting there since the building's east portico redux in 1958. There are open, weedy patches and sometimes a rain pool, and small and large trees that attract butterflies and songbirds. From this vantage point atop the ridge, migrating hawks, loons, swans, and swallows may be seen flying overhead. The bridle trail continues downhill to Rock Creek.

Leaving the Nature Center and stables behind, you can continue south on Glover Road, through the woods, to a ridge top with open grassy areas. Watch for picnic areas 17 and 18. This is another prime warbler-watching spot, best at first light. Birders call this high point on Rock Creek's west bank the Ridge. It is one of the best areas to watch migrating warblers, cuckoos, and the like.

A bit farther down Glover Road, just past the Ross Drive intersection, a large opening and horse corral, known as the Equitation Field (between picnic areas 25 and 26), is another great birding stop during migration periods. Ross Drive is a quiet road running through the park's heart and is worth exploring if time allows.

The scenic Peirce Mill dam in Rock Creek Park. A place to watch for night-herons, wood ducks, and migrating herring along the ladder-like "fishway" installed there.

At these ridge areas, when the first sunlight hits the treetops, the canopy can be alive with tired migrants. Late April to mid-May and late August into October (peak in September) are best.

Trails There are marked trails and many unmarked paths in the park. Some of the park's primary pathways began as old farm roads. Two run north–south through much of the park, from the Maryland line south to just below Peirce Mill. The green-blazed Western Ridge Trail, of moderate difficulty, runs along the park's western ridge for 4.6 miles, connecting near its southern terminus with Melvin C. Hazen Park.

The blue-blazed Valley Trail runs 5.6 miles along the east side of Beach Drive and Rock Creek. It is also of moderate difficulty. Tan-blazed trails connect the two, allowing hikers to make circular routes combining them. These principal trails are maintained by the Potomac Appalachian Trail Club. Hikers may also follow the two equestrian trails and the 9.8-mile bike trail that runs from the Maryland line south to the Arlington Memorial Bridge. Bicycles are only allowed on paved bicycle routes. Park personnel ask that visitors stay on designated trails to minimize erosion and the disturbance to plant life.

Smithsonian's National Zoo South to the Potomac River The National Park Service–administered Rock Creek and Potomac Parkway runs from south of the Smithsonian's National Zoo to West Potomac Park. This 2.5-mile-long scenic drive was built between the 1920s and 1930s to provide a pleasant ride between the monuments and Rock Creek Park. The road's popularity and its need for mainte-

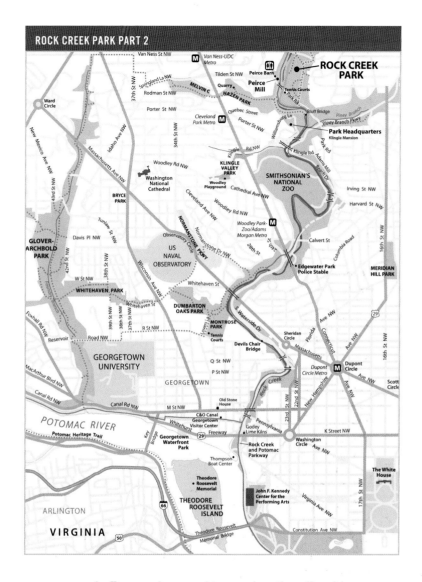

ROCK CREEK PARK PART 2

Van Ness St NW
Van Ness-UDC Metro
Tilden St NW — Peirce Barn
Peirce
ROCK CREEK PARK
Springland La NW
37th St NW
MELVIN C
Quarry
Peirce Mill
HAZEN PARK
Tennis Courts
Rodman St NW
Park Rd
Ward Circle
Porter St NW
Quebec Street
Bluff Bridge
Piney Branch
Idaho Ave NW
34th St NW
Cleveland Park Metro
Porter St NW
Williamsburg Ln
Piney Branch Pkwy
New Mexico Ave NW
Massachusetts Ave NW
WISN St NW
Park Headquarters
Klingle Mansion
Klingle Rd NW
Woodley Rd NW
KLINGLE VALLEY PARK
Klingle Rd NW
Adams Mill Rd
Park Rd
Washington National Cathedral
BRYCE PARK
Woodley Playground
Woodley Rd NW
Cathedral Ave NW
SMITHSONIAN'S NATIONAL ZOO
Irving St NW
Cleveland Ave NW
Harvard St NW
Davis Pl NW
Tunlaw St NW
42nd St NW
38th St NW
Observatory Circle
Woodley Rd NW
Normanstone Dr NW
NORMANSTONE PKWY
Woodley Park-Zoo/Adams Morgan Metro
16th St NW
GLOVER-ARCHBOLD PARK
US NAVAL OBSERVATORY
Wisconsin Ave NW
28th St
Calvert St
Columbia Road
MERIDIAN HILL PARK
W St NW
Whitehaven St
Edgewater Park
Police Stable
WHITEHAVEN PARK
39th St NW
38th St NW
37th St NW
Whitehaven St
DUMBARTON OAKS PARK
S Waterside Dr
Foxhall Rd NW
Reservoir
Road NW
R St NW
MONTROSE PARK
Tennis Courts
Sheridan Circle
Massachusetts Ave NW
Florida Ave NW
Connecticut Ave NW
16th St NW
29
Devils Chair Bridge
GEORGETOWN UNIVERSITY
Q St NW
P St NW
Dupont Circle Metro
Dupont Circle
Scott Circle
GEORGETOWN
New Hampshire Ave NW
20th St NW
22nd St NW
MacArthur Blvd NW
Canal Rd NW
M St NW
Old Stone House
Rock Creek
Canal Rd NW
C&O Canal
Georgetown Visitor Center
Whitehurst Freeway
Pennsylvania
Godey Lime Kilns
K Street NW
Washington Circle
Ave NW
POTOMAC RIVER
Potomac Heritage Trail
Key Bridge
Georgetown Waterfront Park
29
Rock Creek and Potomac Parkway
23rd St NW
Thompson Boat Center
The White House
ARLINGTON
66
Theodore Roosevelt Memorial
John F. Kennedy Center for the Performing Arts
Virginia Ave NW
17th St NW
THEODORE ROOSEVELT ISLAND
VIRGINIA
50
Theodore Roosevelt Memorial Bridge
Constitution Ave NW

nance spurred efforts to clean up this once heavily polluted lower part of the park. Roadway aficionados call it a fine example of early motor parkway design.

Other Highlights of Rock Creek Park While much of the park's history lies hidden in the woods, Peirce Mill is an exception. Tours are available and other park programs start there. Peirce Mill was the last of many water-powered mills that local farmers relied on from the late eighteenth to early twentieth centuries. In 2006, a fish ladder was built so herring could migrate above Peirce Mill falls to reach spawning areas farther upstream. The best time to see these fish is usually late April into early May.

Fort DeRussy's remains lie in the woods off the bike path, just off Oregon Avenue, N.W. close to where this street intersects Military Road, N.W.

A 1.5-mile exercise course starts near Calvert Street, N.W., and Connecticut Avenue, N.W., and runs along the upper Rock Creek and Potomac Parkway.

Rock Creek Park Horse Center offers riding lessons and guided trail rides on the park's 13 miles of bridle trails.

Twenty-five tennis courts are available at 16th and Kennedy Streets, N.W. (Fees and reservations are required, except from late fall to early spring.)

Rock Creek Golf Course is an 18-hole public golf course, reached from 16th and Rittenhouse Streets, N.W. Rentals are available and a fee is charged. Phone (202) 882-7332.

Carter Barron Amphitheatre is a 3,700-seat outdoor theater, where concerts take place in warm months. The box office is located at 16th Street and Colorado Avenue, N.W. Phone (202) 426-0486.

Picnicking is permitted at 29 picnic areas; some require reservations.

The Planetarium is located at the Nature Center, gives programs focused on the skies over Washington, D.C.

Smithsonian's National Zoo is nestled in a bend of Rock Creek and is adjacent to the park (see separate entry).

Thompson Boat Center is located at the very bottom of the park where Rock Creek drains into the Potomac River and where the Rock Creek and Potomac Parkway meets Virginia Avenue, N.W. Roosevelt Island (see separate entry) is visible from the docks, directly across the Potomac River. Here you can rent kayaks, canoes, small sailboats, rowing shells, and bicycles. Boating lessons are available. (Parking is limited; additional parking is available at nearby parking garages.) Phone (202) 333-4861.

Georgetown Waterfront Park sits just west, a grassy area on the banks of the Potomac. Parking is available on K Street, N.W., under the Whitehurst Freeway.

GETTING THERE

By Metrorail Four Red Line Metrorail stations are located near Rock Creek Park. From the Silver Spring station in Maryland, you can walk west on Colesville Road to reach the north end of the park. It turns into North and South Portal Drive, N.W., before intersecting with East Beach Drive, which in short order will take you south into the park.

The Van Ness station is located on Connecticut Avenue, N.W., just south of Soapstone Valley Park. A path starting just east of Connecticut Avenue, off Albermarle Street, N.W., heads east through this park and joins Rock Creek Park a bit north of Peirce Mill.

The Cleveland Park station is also located on Connecticut Avenue, N.W. Just north of this station, a trail leaves from the east side of Connecticut Avenue, heading down a ridge into Melvin C. Hazen Park and continuing east into Rock Creek Park.

The Woodley Park-Zoo / Adams Morgan station accesses the park's southern section. Immediately south of the station 24th Street, N.W., forks to the right off Connecticut Avenue, N.W., leading to the Rock Creek and Potomac Parkway (just south of the Smithsonian's National Zoo).

By Metrobus Buses D31, D33, D34, E2, E3, E4, and W45 stop at or near the Glover Road / Military Road intersection, from where you can walk south on Glover Road into the park, near the Nature Center. (See MetroOpensDoors.com to confirm routes or to find other ones near other points in the park.)

By Car The Military Road, N.W., entrance takes visitors south on Glover Road and into the park near the Nature Center.

Beach Drive runs south from the Maryland line, and there are many places to park and explore. To reach the Nature Center from Beach Drive, take Military Road, N.W., westbound and then turn left onto Glover Road.

Note: During weekends (7 a.m. Sat. to 7 p.m. Sun.) and holidays, the following sections are closed to car traffic: Beach Drive between Military and Broad Branch Roads; from the Maryland state line south on West Beach Drive; and Wise Road south to Picnic Grove 10. During these times, cyclists, runners, and walkers rule these roads.

By Bike The Rock Creek Trail, a signposted asphalt bike route, runs for 25 miles, mostly through wooded parkland. Starting from the Lincoln Memorial, it heads north through Rock Creek Park, into Maryland and through Rock Creek Stream Valley Park, and then into Rock Creek Regional Park's Lake Needwood, just north of Rockville, Maryland.

Rock Creek Park is one of the nation's oldest national parks and one of its largest and wildest city parks.

At the Rock Creek Trail's south end, the Memorial Bridge (Arlington Memorial Bridge) connects this route to the Mount Vernon Trial in Virginia.

During weekends (7 a.m. Sat. to 7 p.m. Sun.) and holidays, the following sections are closed to car traffic and open to cyclists: Beach Drive between Military and Broad Branch Roads; from the Maryland state line south on West Beach Drive; and Wise Road south to Picnic Grove 10.

NEARBY

Fort Bayard Park   Picnic sites and a ball field now stand at this location on River Road at Western Avenue, N.W., once part of a ring of Civil War forts protecting the city.

Fort Reno Park   Little remains of this fort, once called Fort Pennsylvania, which commanded the highest point in the city, at 429 feet. In July 1864, Union forces' long-range cannons fired from Fort Reno and three and a half miles to the north killed four Confederate troops at the site of what is now the Walter Reed National Military Medical Center in Bethesda, Maryland.

Fort Stevens Park   Located just east of Rock Creek Park at 13th and Quackenbos Streets, N.W., this is the area where the only Confederate attack on the city took place. The guns at Fort Reno, located west, were not used to defend this fort because Union forces were afraid Fort Reno's long-range cannons would accidentally kill nearby Union troops.

Battleground National Cemetery   Just to the north, off Georgia Avenue, N.W., sits the country's second-smallest national cemetery. Forty-one soldiers who died defending Fort Stevens are buried here.

Melvin C. Hazen Park   This narrow park joins Rock Creek Park just below Peirce Mill. Parking is available off Williamsburg Lane, N.W., off Porter Street, N.W. From Klingle Mansion—Rock Creek Park's headquarters—trails lead north to Peirce Mill and west to Connecticut Avenue, N.W.

Smithsonian's National Zoo

LOCATION
3001 Connecticut Avenue, N.W.
Washington, DC 20008

TELEPHONE
National Zoo Information Line: (202) 633-4800

WEBSITE
http://nationalzoo.si.edu/

SIZE
163 acres

HABITATS
mature hardwood forests and ridges, landscaped gardens, artificial ponds, Rock Creek, streamside woodland and brush

Natural History   Occupying picturesque hillsides and part of the Rock Creek Valley, the National Zoo is a haven not only for zoo animals but also for local wildlife. The zoo sits at a sharp bend of snaking Rock Creek, which over millions of years slowly wore its way through the area's resistant metamorphic rock, cutting terraces that are now covered in thick vegetation. Hardwood trees cloaking the ridges here are among the oldest in the region and include northern red, eastern white and chestnut oaks, tuliptree, American beech, red maple, and American sycamore. The rich soils and thick leaf litter beneath these giants nurture Christmas and lady ferns, ebony spleenwort, and shrubs and small trees such as spicebush, pawpaw, and flowering dogwood. This rolling landscape falls within the geographic province known as the Piedmont. Not far to the east, around 16th Street, the Piedmont meets the Coastal Plain, a province known for flatter terrain and evenly worn sediments.

Many of the zoo's buildings contain locally quarried stone. The walls of the Panda House and some other buildings, for example, contain blocks of Potomac bluestone, which is more than 500 million years old. About the same age, Kensington gneiss adorns parts of the Think Tank exhibit and parts of the American Trail.

Human History   Native Americans once hunted and fished along Rock Creek here. Although a water-powered mill stood on part of the property, most of the acreage was unsettled in March 1889, when Congress and President Grover Cleveland signed off on the land's acquisition and the zoo's creation. Interest in creating a large zoo in the nation's capital had sprung from a wildly popular live animal exhibit started on the National Mall in 1887. Civic leaders,

This clock adorns the zoo's Lion/Tiger Hill, near Beach Drive.

Giant pandas share their large, leafy outdoor enclosures with native wildlife, including eastern chipmunks and nesting songbirds.

including banker Charles Carroll Glover, championed the zoo's creation.

Today, many visitors come to the National Zoo to see giant pandas, Asian elephants, sloth bears, orangutans, and tigers. However, gardeners and local wildlife aficionados also find much of interest here. Famed landscape architect Frederick Law Olmsted, who previously put his genius to work designing New York City's Central and Prospect Parks, drew up a plan for a winding walk that traversed the zoo's striking natural features, blending animal exhibits with the beautiful setting. The zoo's main path is named in his honor.

Olmsted Walk   Olmsted Walk runs past many of the zoo's top exhibit areas, beginning from the top of the hill at Connecticut Avenue, N.W., and the Visitor Center and ending at the bottom by Rock Creek. Along the way, the landscaping and, in spots, mature hardwood forest border the path. By 9 or 10 a.m. on most mild or warm days, the zoo is busy with foot traffic and parking lots quickly fill. To see the zoo's native wildlife, arrive as early as possible. Eastern chipmunks, gray squirrels, gray catbirds, cardinals, and other species forage near the main walkways. During migration, almost any eastern woodland bird may show up, while overhead, hawks and vultures—even on occasion a bald eagle—may take advantage of thermal air currents rising above the valley. Fish crows regularly fly over, their "uh-uh" calls contrasting with the familiar "caw-caw" of the American crow.

Gardens, ponds, and fountains dot the walk, many designed with wildlife in mind. Some areas are stocked with plants that draw ruby-throated hummingbirds, butterflies, and bees. In spring and summer, tiger swallowtails, skippers, and monarchs are among the commonly seen butterflies. Local reptiles sometimes appear in the rich garden soils or nearby woodland, including northern brown snake. Berry-bearing trees such as American holly and shrubs, including winter berry, attract cedar waxwings, American robins, northern mockingbirds, and gray catbirds. In summer and early fall, gaudy American goldfinches perch atop purple coneflowers and black-eyed Susans, busily extracting their seeds.

Around Lemur Island, about halfway down the walk, watch the treetops for Baltimore and orchard orioles and eastern kingbirds, which nest there.

Asia Trail and Elephant Areas   Over the past 10 years, large parts of the zoo, particularly the Asia Trail, elephant areas, and American Trail, have been transformed. Some mature trees fell in the process; many new areas were planted. Concerted efforts were taken to leave some mature trees, building and planting around them. The result is a mixed mosaic of wild forest patches along ridges and diverse landscaping along main exhibit areas. This varied vegetation attracts wild birds year round. For example, Asia Trail's sloth bear exhibit, near the Visitor Center and the Connecticut Avenue entrance, is where visitors may watch sloth bears share the

same real estate with migrating thrushes and wintering white-throated sparrows. In spring and summer, ruby-throated humming-birds may dart between gardens near Asian small-clawed otters or clouded leopards.

Rock Creek at the Zoo Like much of adjacent Rock Creek Park, the forest here is older than most forest elsewhere in the metropolitan area. The easiest access to the creek is from the bottom of the hill, near parking lots D and E. A stretch of bike path runs by a bridge near the Amazonia exhibit and affords nice views of the creek and adjacent brush and forest. Wood ducks, mallards, and belted kingfishers are frequently seen in the vicinity, sometimes joined by stalking black-crowned night-herons and great blue herons. Night-herons nest at the zoo (see "The Bird House" section).

Summering forest birds that nest and sing along Rock Creek include red-eyed vireo, Acadian and great crested flycatchers, and blue-gray gnatcatcher. Nesting resident birds include red-shouldered hawk, barred owl, and hairy and pileated woodpeckers. Fall, winter, and early spring visitors include Cooper's and sharp-shinned hawks, yellow-bellied sapsucker, brown creeper, winter wren, hermit thrush, and a variety of sparrows. White-tailed deer are common in adjacent Rock Creek Park and may forage at the zoo. Red foxes and coyotes are in the area but are rarely seen by visitors.

The Bird House Home to more than 100 species from around the world, the zoo's Bird House also attracts many wild birds. The back of the Bird House and its east-facing side overlook a ridge of mature forest. There, birders can spend delightful spring and fall mornings watching migrant songbirds at eye level in the treetops, while also stealing glimpses of several of the zoo's crane species, marabou storks, and kori bustards. The Bird House's west side connects to the Great Flight Exhibit, a large enclosure that has housed a variety of birds since 1965.

A boardwalk and network of man-made ponds face the Bird House's front entrance. This artificial wetland provides habitat for bullfrogs and a few turtles. Any ducks seen here are wild, including mallards and some wood ducks, a few of which may be present through the winter. On occasion, an American black duck or teal shows up. The river birches and other vegetation flanking these ponds may attract migrating warblers, vireos, and orioles, or wintering kinglets, among other songbirds.

A flamingo flock lives behind the Bird House in an enclosure that often attracts wild wood ducks, mallards, and sometimes black-crowned night-herons. The trees around the Bird House and directly across the path from the flamingos usually hold dozens of black-crowned night-heron nests, most of which are hidden beneath the leafy canopy. Members of this long-standing wild colony return in the first days of March. Nesting activity, though, kicks into full swing in early April and continues through summer, when white-spotted young birds loiter around the Bird House's pools and Rock Creek. This is the most reliable spot in the city to see this species.

American Trail This trail runs along the bottom of the Rock Creek Valley, below the Bird House and elephant areas. Part of it is landscaped to resemble the Pacific Northwest and part immerses you in mature native deciduous forest. In the enclosures, you will see Mexican wolves, river otters, and seals and sea lions. The stunning Amazonia exhibit and Kids' Farm are here as well. Early in the morning, you may see wild white-tailed deer, gray squirrels, eastern cottontail rabbits, eastern chipmunks, and woodpeckers here.

GETTING THERE

By Metrorail Two Red Line stations are a 15-minute walk from the zoo: Cleveland Park and Woodley Park-Zoo / Adams Morgan. First-time visitors often choose the latter because of its name, but the walk to and from Cleveland Park is level, while that from Woodley Park-Zoo / Adams Morgan to the zoo is uphill. So, walk from the Cleveland Park station to the zoo and then, later, walk downhill to the Woodley Park-Zoo / Adams Morgan station.

Olmsted Walk, the zoo's main path, is eight-tenths of a mile long and slopes down toward the creek. At the end of the trip, Metrorail riders will need to walk back up the hill to exit toward the stations. A day at the zoo, including roundtrip walks to the Metrorail, is a good workout.

By Metrobus Four Metrobus lines stop at the National Zoo. Lines L1, L2, and L4 stop at the Connecticut Avenue, N.W., entrance, near the Visitor Center. The H2 stops at the Harvard Street entrance, at the bottom of the hill.

By Car The main entrance to the National Zoo is located at 3001 Connecticut Avenue, N.W. This is at the top of the zoo's formidable hill. There are also two entrances on the zoo's east end, at the bottom of the hill: one off Beach Drive (just north of where it changes name to Rock Creek and Potomac Parkway) and the other at Harvard Street.

Parking, however, is very limited, especially on weekends and any day with nice weather. If you drive, arrive early; before 9 a.m. is best. There are five lots: Lot A is at the top of the hill near Connecticut Avenue and the Visitor Center. Lot B, the smallest, is near the Elephant Community Center. Lot C is an overflow lot and is sometimes closed. Lots D and E are at the bottom of the hill and often fill up last. Zoo admission is free, but parking will cost you.

By Foot There are three main foot and vehicle entrances. See "By Car" section. *Note*: On Connecticut Avenue, N.W., are separate car and pedestrian entrances. Pedestrians enter via the open gates and wide entrance that marks the beginning of Olmsted Walk (just south of the vehicle entrance).

Once in the park, the only way to get around is by foot. Visitors with children under three or four years old may want to bring a

Wild black-crowned night-herons once nested on the outside of the zoo's Great Flight Exhibit. Today, the birds nest in nearby trees.

stroller or rent one at the zoo. The zoo's concessions sell food and drinks, but visitors may also bring their own.

Visitors not accustomed to long walks will want to pace themselves, particularly during the hot, muggy summer. Hats, sunscreen, and water are recommended because although lined with trees most of the way, Olmsted Walk gets little shade.

The grounds open early, but the zoo buildings don't open until 10 a.m. The zoo is open every day of the year, except December 25. During harsh winter weather, it may also close. Admission is free.

NEARBY

Meridian Hill Park　Located east of 16th Street, N.W., just east of the zoo's south end, this narrow park with century-old oaks attracts an array of spring and fall migrants.

McMillan Reservoir　East of the zoo and adjacent to Howard University, McMillan Reservoir attracts ducks, grebes, gulls, and a heron or two. The reservoir can be viewed from the sidewalk through the fence surrounding it.

Rock Creek Cemetery　Rock Creek Cemetery is one of the city's oldest cemeteries, dotted with old oaks and other vegetation. This property is open to the public. It can be an important stopover for migrating birds. Eastern wood-pewee, great crested flycatcher, and orioles nest here. This property is near the Fort Totten station on the Metrorail's Red Line.

Fort Totten Park　Like a densely forested teardrop surrounded by a heavy grid of streets, this Northeast Washington, D.C., park attracts migrating songbirds. Metrorail's Red Line station at Fort Totten is nearby.

Glover-Archbold Park

LOCATION

Stretching from the C&O Canal National Historical Park boundary north to the corner of Van Ness Street, N.W., and Wisconsin Avenue, N.W., this park sits between two other north–south ribbons of green in this part of town, with the much larger Rock Creek Park to its east and the smaller Battery Kemble Park to the west.

TELEPHONE

Contact the Rock Creek Park Nature Center at (202) 895-6070.

SIZE

183 acres and about three miles long

HABITATS

stream valley covered with hardwood forest

Natural History The tall trees can make wildlife viewing a challenge, but the forest canopy guarantees a feeling of seclusion for anyone entering the park and following the trails there. Beneath the canopy grow American holly, spicebush, and other small trees and shrubs typical of the area. In spring, an invasive nonnative wild-flower called fig buttercup cloaks many areas in green and yellow. Along the stream, American sycamore and tuliptree are among the common trees, while drier ridges along the sides of the park include American beech and northern red and eastern white oaks. Common wildlife includes white-tailed deer, gray squirrel, northern raccoon, and a variety of woodland birds, including barred and eastern screech-owls, pileated woodpecker, and red-shouldered hawk. Glover-Archbold is a good place to witness migration, when mornings are often brightened by first sightings of such overnight arrivals as warblers, grosbeaks, orioles, and tanagers. Winter birds include winter wren and flocks of white-throated sparrows and juncos.

The shady woodland is important for the city's official bird, the wood thrush. Its flute-like song rises at the end. Another nesting thrush species, the veery, sings a distinctive downward-spiraling song that has an ethereal, almost air-chamber-like quality. In the city, the veery is at the very edge of its breeding range.

Until the 1940s, naturalists thought veeries only nested in the Appalachian Mountains and areas farther to the north. In May 1942, U.S. State Department employee and amateur naturalist Louis J. Halle found a pair nesting in a ravine in the lower part of Rock Creek Park. In 1943, his account of this first known D.C. record for veery nesting appeared in the ornithological journal *The Auk.* Halle later

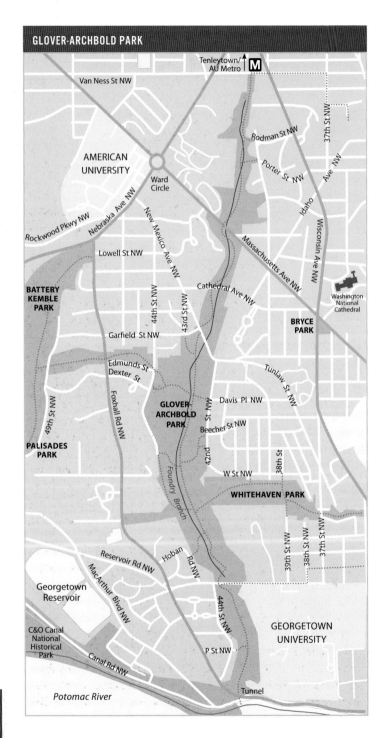

Tenleytown/
AU Metro **M**

Van Ness St NW

Rodman St NW

37th St NW

Porter St NW

Idaho Ave NW

AMERICAN
UNIVERSITY

Ward
Circle

Rockwood Pkwy NW

Nebraska Ave NW

New Mexico Ave NW

Massachusetts Ave NW

Wisconsin Ave NW

Lowell St NW

BATTERY
KEMBLE
PARK

Cathedral Ave NW

Washington
National
Cathedral

44th St NW

43rd St NW

BRYCE
PARK

Garfield St NW

Edmunds St
Dexter St

Foxhall Rd NW

Tunlaw St NW

49th St NW

GLOVER-
ARCHBOLD
PARK

St NW

Davis Pl NW

Beecher St NW

38th St

PALISADES
PARK

42nd

W St NW

Foundry Branch

WHITEHAVEN PARK

39th St NW

38th St NW

37th St NW

Reservoir Rd NW

Hoban Rd NW

Georgetown
Reservoir

MacArthur Blvd NW

44th St NW

GEORGETOWN
UNIVERSITY

C&O Canal
National
Historical
Park

Canal Rd NW

P St NW

Tunnel

Potomac River

showed these birds to another local naturalist, *Silent Spring* author Rachel Carson.

Watch for roosting barred owls in Glover-Archbold Park's tall trees.

Human History In 1924, realizing that people needed escape from the urban landscape, banker and parks advocate Charles Carroll Glover joined forces with Anne Archbold, champion of women's rights and the outdoors and daughter of oil millionaire John D. Archbold. The two donated 100 acres to the public as a nature preserve and escape for urbanites that today bears their names.

Well-connected and passionate about his city, Glover was also an important driver for projects that now define the D.C. landscape. These include the Washington National Cathedral, the Washington Monument (he pushed for its completion), and some of the parks described in this book. He arranged land purchases for the government but also donated acreage. His participation was crucial in the establishment of not only Glover-Archbold Park but also the National Zoo, Rock Creek Park, Rock Creek and Potomac Parkway, the development of East and West Potomac Parks, and the layout of the upper part of the city's Northwest quadrant.

The donated Glover and Archbold properties had a right-of-way running through them, and the city planned to build a parkway. After World War II, Archbold fought the growing momentum to build the road. In 1967, the year before she died, the D.C. government gave up the right-of-way, and the park has been a safe haven for recreation and nature ever since.

The National Park Service administers this park, but unlike nearby Rock Creek Park, it has fewer facilities.

Trails Glover-Archbold Park is treasured by local residents and virtually unknown to many others. Though the park lacks infrastructure, it has marked trails. Once you are inside, you can hike without fear of getting lost because the park is long but narrow, with a stream running through the middle.

The Glover-Archbold Trail is the park's main trail. This wide pathway follows the Foundry Branch stream for 3.1 miles, the length of the park. This trail can be especially productive for birding from where Reservoir Road crosses the park north for at least a mile. The trail is simple, in spots crossed with roots, and it can be muddy. The

Glover-Archbold Trail's northern terminus is Van Ness Street, N.W., in the Tenleytown neighborhood.

Most of the trail passes through woodland but just south of Reservoir Road, and adjacent to Georgetown University, there is an open area with picnic tables. The trail re-enters the forest before it ends to the south just below Canal Road, at the C&O Canal (see separate entry) and Potomac River. At its end, the trail goes under an old trestle for the trolley that once connected Georgetown with the amusement park at Glen Echo, Maryland. Then it passes through a short tunnel under the C&O Canal and opens at the towpath.

Whitehaven and Dumbarton Oaks (see separate entry) Parks connect Glover-Archbold with Rock Creek Park, which lies to the east. From inside Glover-Archbold, you can take the Whitehaven Trail east 0.9 miles through Whitehaven Park to Wisconsin Avenue, N.W. If you want to continue east, cross Wisconsin Avenue and the trail continues, passing through Dumbarton Oaks Park, to Montrose Park (see separate entry). In Montrose Park, a nearby trail loops into Rock Creek Park. From Glover-Archbold Park to Rock Creek Park via this trail is a hike of about 1.5 miles.

The Wesley Heights Trail connects to Battery Kemble Park and Palisades Park to the west. The trail starts from 42nd Street, N.W., at the east edge of Glover-Archbold Park and then crosses the park on the other side, crossing 44th Street, N.W., and then 49th Street, N.W., before entering Palisades Park. Battery Kemble Park, connected to Palisades Park at its north end, is another recommended spot for birding during migration periods.

Serious hikers can stitch these together to make a 5.5-mile loop connecting Glover-Archbold's Wesley Heights Trail with trails in Palisades Park to the west and the C&O Canal's towpath to the southwest.

Before embarking upon one of these longer, park-connecting hikes, you can direct any questions to the Rock Creek Park nature center staff at (202) 895-6070.

GETTING THERE

By Car From Wisconsin Avenue northbound, turn left onto Cathedral Avenue and then left onto New Mexico Avenue, N.W. Park on New Mexico Avenue or 42nd Street. There is a marked trail entrance located near the intersection of Garfield Street, N.W., and New Mexico Avenue, N.W.

In a busy part of the city, yet away from it all.

In its northern section, the park is crossed by New Mexico Avenue and Garfield Street, Cathedral Avenue, Massachusetts Avenue, and Van Ness Street, at its narrow top. Part of the park is flanked on the west by 44th Street and on the right by 42nd Street. To avoid being ticketed, park only where signs indicate regular parking is permitted.

For access to the southern part of the park, there are metered spots off Reservoir Road, between Georgetown University and the French Embassy and Foxhall Road.

By Metrorail   The Red Line Tenleytown/AU station is the closest Metrorail station to the park, just a few blocks north of the park's narrow northern entrance at Van Ness Street. Walk south on Wisconsin Avenue, N.W., and turn right on Van Ness Street, N.W., to see the park entrance on the left side of the street.

This entrance can also be reached from the Red Line Van Ness station by walking almost a mile west on Van Ness Street, N.W. These routes, however, do not give easy access to the widest part of the park.

By Metrobus   A number of bus routes pass by or through the park; some of them are linked to Metrorail stations, such as the Red Line Dupont Circle station. Check MetroOpensDoors.com for current routes. For example, from the Red Line Dupont Circle Metro station, the N2 and N6 buses run northwest and cross the park on Cathedral Avenue, N.W., near the north end of the park.

Community gardens highlight local residents' love for the park and their neighborhood.

The D1 bus accesses the park's midsection. From the west end of Calvert Street, at 41st Street, N.W., a rider need only walk a block west to 42nd Street, N.W., to reach the park.

Along the park's south section, the D6 bus runs from the Red Line Dupont Circle Metro west along Reservoir Road. The park's north–south Glover-Archbold Trail crosses Reservoir Road, N.W., on the park's west side by the intersection with 44th Street, N.W.

NEARBY

Battery Kemble Park   Bounded to the west by Chain Bridge Road and the east by 49th Street, N.W., this narrow, mile-long park runs north–south and is a hidden gem that protects a Civil War fort site that overlooked approaches to Chain Bridge. Today, walking the slopes is a green escape from the traffic and a great place to be during spring and fall migrations. Areas of meadow and groves of Virginia pine give this small park varied terrain for seeking butterflies, wildflowers, birds, and other wildlife. In fall and winter, the pines may harbor owls and interesting winter songbirds. Enter the park via trails accessible from the intersection of Loughboro Road / Nebraska Avenue, N.W., and Foxhall Road, N.W., to the north and MacArthur Boulevard, N.W., and Chain Bridge Road, N.W., to the south. The park connects to Glover-Archbold Park via a green strip located below Fulton Street. A car entrance to the park is on the east side of Chain Bridge Road, N.W., one-quarter mile south of Loughboro Road, N.W.

Chesapeake and Ohio Canal
National Historical Park

LOCATION

The towpath starts in Georgetown and runs about 3.5 miles north-west to the Maryland line. (In Maryland, it continues for 181 more miles.) The Washington, D.C., visitor center is in Georgetown at 1057 Thomas Jefferson St., N.W.

TELEPHONE

(202) 653-5190

WEBSITE

http://www.nps.gov/choh

SIZE

The entire park spans 19,236 acres and is 184.5 miles long, from Georgetown to the western Maryland city of Cumberland. Only a few miles fall within Washington, D.C.

HABITATS

river, rocky and muddy shoreline, floodplain forest, brushy patches, and maintained and unmaintained canal sections

Natural History The C&O Canal is a man-made aquatic waterway paralleling the Chesapeake Bay's second-largest tributary, the Potomac River. Like the Potomac it parallels, the canal passes through diverse wildlife habitat and four physiographic provinces. At the mouth of Rock Creek, where the canal begins, Coastal Plain yields to Piedmont. Farther west in Maryland, the canal passes through two mountainous provinces, the Blue Ridge and the Valley and Ridge, before ending in Cumberland, Maryland, at the start of the Allegheny Plateau.

A snaking, river-edge ribbon of park, the C&O Canal National Historical Park provides an important corridor for plant and animal dispersal in an otherwise fragmented landscape. Floodplain forest composes more than 80 percent of the property. There, nutrient-rich silt laid down by heavy river flows supports rich floral diversity, where ranges for many northern and southern species overlap. More than 2,000 species of plant and animal have been identified within the park.

From late March into May, spring wildflowers blanketing the muddy soil include mayapple, Virginia bluebells, spring beauty, wild blue phlox, and Dutchman's breeches. Large riverside trees tower overhead. These include American sycamore, tuliptree,

The mule teams and barges are history. Today, the C&O Canal is a major recreational corridor.

boxelder, and silver maple. Beneath the canopy flourish stands of pawpaw and spicebush. Much of the forest you now see along the Potomac River and the C&O Canal that parallels it grew back following the canal's closure in 1924. Invasive nonnative plants such as fig buttercup, Japanese stiltgrass, and Japanese honeysuckle now dominate the undergrowth in some parts.

Beaver, white-tailed deer, cottontail rabbit, opossum, and raccoon are among the canal's common mammals. But often the birds are far easier to see. One of the city's top birding locations, the C&O Canal offers something to see in each season, including orioles, warblers, swallows, woodpeckers, gulls, wood ducks and other waterfowl, great blue herons, double-crested cormorants, hawks, barred owls, ospreys, and bald eagles. (For years, a bald eagle pair has nested north of the city along the Maryland towpath on Conn Island, which can be viewed from near the parking lots at Great Falls, Maryland.) More than 100 bird species have nested in the park. Although many of these were recorded in the wilder Maryland portions, the D.C. portion receives most of these species during migration.

Human History For many decades, Native Americans and early colonists traded along the Potomac River. In 1608, John Smith took a ship up the Potomac as far as Little Falls, near the current boundary line between the district and Maryland. The rapids kept him from traveling farther upstream. In the 1600s, the colonies of Virginia, Maryland, and Pennsylvania were established. After inde-

CHESAPEAKE AND OHIO CANAL
NATIONAL HISTORICAL PARK

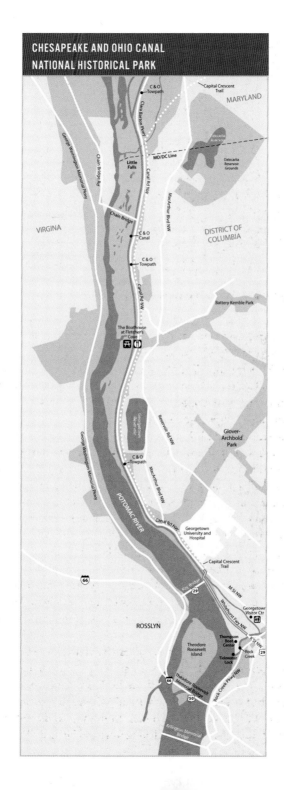

pendence and as the country grew, it became clear that a connection should be made between the Potomac and the rapidly growing Ohio Valley to the west. Starting in 1785, the Patowmack Company, at first presided over by George Washington, began working on a series of canals that skirted Little Falls and four other rapids on the Potomac. Washington became the first president in 1789, selected the site for the new capital, and had many other distractions, but he kept an interest in the canal's construction. In 1802, three years after his death, construction was completed on the last locks, at Great Falls. For about 25 years, the canal system enabled boats to travel the river, circumventing troublesome rapids. But people started to realize that transport would be much easier if it all took place alongside the temperamental Potomac.

In 1828, the Chesapeake and Ohio Canal Company purchased the aging Patowmack Canal with hopes of building a complete "water highway" alongside the Potomac, one that would reach the Ohio River and Pittsburgh, Pennsylvania. The Baltimore & Ohio (B&O) Railroad broke ground the same year.

Because of 74 locks, the C&O Canal rose 605 feet along its 185-mile route from Georgetown, just west of the Fall Line, to Cumberland in Maryland's western mountains. Engineers designed the locks to raise or lower boats as needed. Water was drawn in from the Potomac via feeder dams, and culverts diverted excess canal water back into the river or, in the city, to the water-powered machinery of Georgetown factories. Ninety-foot-long canal boats slowly navigated the canal, pulled by mule teams that clopped along the towpath at up to four miles per hour. At the canal's peak in the 1870s, about 2,000 mules pulled some 800 boats along the canal. Some of the original structures, including lock houses and refurbished locks, give the visitor a feel for this nineteenth-century canal era, particularly at Georgetown and Great Falls, Maryland. Restored canal boats may be seen at Georgetown and Great Falls, Maryland.

By the time it opened for business in 1850, the canal was losing business to the railroad, which had already pushed farther west. In the end, the canal never reached Pennsylvania. It ended in Cumberland, Maryland, which the railroad had reached eight years earlier. But until after the Civil War, the canal had one advantage: It was the only direct route linking coal-rich Cumberland to the capital. During the Civil War, Confederate and Union forces crisscrossed the canal en route to battles such as Antietam and Gettysburg. Confed-

The canal parallels the Potomac, and both are flanked by a miles-long ribbon of forest that provides nature enthusiasts with endless hours of enjoyment.

erate forces targeted canal commerce in their attempts to cut supply lines to Washington.

Just after the war, the canal had its most productive years, but by the 1880s, business had slipped. Railroads chugged ahead. In an ironic twist, the B&O Railroad saved the canal from ruin by buying it after a crippling 1889 flood. The canal operated until 1924, when another flood caused serious damage.

Shortly before the Great Depression, there was talk of turning the run-down canal and towpath into a highway. But in 1938, the federal government bought it from the railroad for $2 million. It became a project for the job-generating Civilian Conservation Corps, which restored the first 20 miles of the canal. A 1942 flood washed much of this work away.

By this time, naturalists and walkers appreciated the natural virtues of the crest-fallen canal. Birding and field guide pioneer Roger Tory Peterson lived in the city for 10 years during and after World War II and wrote, "Where is there a walk more lovely in the month of May than the towpath along the old Chesapeake and Ohio Canal?"

After World War II, plans for dams and highways threatened to change the towpath and canal forever. In 1954, Supreme Court Justice William O. Douglas fought road and dam plans, spearheading an eight-day hike that took interested parties the length of the towpath. Debate continued until 1971, when the canal was designated a National Historical Park. Occasional large floods, including two in 1996, provide great challenges for the National Park Service, as it tries to preserve history and keep the canal and Potomac shore safe for the public and productive for wildlife.

The Towpath   Once a wide trail walked by guides and mule teams pulling boats up and down the canal, the towpath is now a great way to see some of the city's most interesting wildlife. A key component of the city's hiker/biker trail network, the Capital Crescent Trail joins the towpath for much of its length in D.C. (Eleven miles long, the Capital Crescent Trail follows the route of the former Georgetown Branch of the Baltimore & Ohio Railroad, connecting Georgetown with Silver Spring, Maryland.)

The towpath can be especially busy on weekends and holidays. In the past, mule teams kept the speed limit on the canal to 4 miles per hour. These days, 15 miles per hour is the towpath speed limit, so hikers need to watch for bikers and stay to the right side of the path, particularly near Georgetown. In general, early morning and week-day walks tend to be much more tranquil.

The C&O Canal officially starts at the mouth of Rock Creek at the Tidewater Lock and it runs to the Maryland line, 4.7 miles to the northwest. The now-decrepit structure inspired the name of the much more recent and now famous Watergate complex, which looms over it today. Long before Nixon, the Tidewater Lock allowed canal boats to turn and enter the canal from the river. Right by the lock, the Thompson Boat Center rents bicycles, canoes, and boats (http://www.thompsonboatcenter.com/).

The towpath starts close by—a short walk west of the Tidewater Lock in trendy, historic Georgetown. Stroll west on this level path, and within a mile, you leave behind the streets, watered sod, potted plants, and restored buildings that once housed factories and ware-houses. The floodplain forest that immerses you continues to Cumberland, Maryland, almost two hundred miles away in the Appalachians of western Maryland.

When walking the towpath, don't forget to stop and scan where river views are best. One favorite area for wildlife fans is 3.2 miles northwest of the Tidewater Lock, at the Boathouse at Fletcher's Cove. As at other canal spots, smallmouth and largemouth bass, white and yellow perch, channel and blue catfish, bluegill, pump-kinseed, and carp frequently take the bait. Many other fish show up here. Recently, anglers began catching Asiatic fish called northern snakehead. The introduced snakeheads concern conservationists

Hooded mergansers are among the many ducks seen in the park.

because these fish prey on a wide variety of other fishes. One reason anglers love this stretch of the Potomac is because a barrier at nearby Little Falls forces fish into a narrow channel. Each spring, herring, striped bass (rockfish), and shad heading upriver to spawn concentrate there.

From early March through October, bike, canoe, kayak, and rowboat rentals are available at the Boathouse at Fletcher's Cove, as is fishing information on which species are caught in which season, fishing supplies and licenses, and refreshments. You will also find picnic tables and restrooms there. For more information, see http://www.fletcherscove.com/

Birders also flock to Fletcher's Cove and the rest of the towpath to see a seasonally changing variety of land and water birds. Nesting songbirds include orchard and Baltimore orioles, warbling vireos, and eastern kingbirds. Late April to mid-May, this area is one of the best places in the city to listen and look for prothonotary and yellow-throated warblers. In spring and fall migrations, swallows, swifts, gulls, Caspian terns, and spotted and solitary sandpipers frequently pass over the Potomac or rest along its shores. As the cold sets in, diving ducks such as bufflehead arrive from northern haunts and show up on the Potomac.

A mile north of Fletcher's Cove, towpath hikers can access the Chain Bridge to get a view looking north at the Potomac and adjacent rocks and ponds that attract herons, migrating snipe, and ducks. Ospreys, vultures, hawks, and occasionally a bald eagle may sail by. From this point, it's a half mile north on the towpath to the Maryland line.

By Metrorail   Visitors using Metrorail can reach the start of the towpath at Georgetown from the Foggy Bottom station on the Blue and Orange Lines. Head north from the station exit on 23rd Street, N.W., west on Pennsylvania Avenue, N.W., and then to M Street, N.W., westbound. Turn left (south) on 30th Street, N.W., to the canal.

To reach the Boathouse at Fletcher's Cove from the Dupont Circle station on the Red Line, take the D6 bus west to the far end of Georgetown Reservoir on MacArthur Boulevard, N.W. Cross the street and go left, following Reservoir Road, N.W., about a quarter mile to the boathouse. Use caution crossing very busy Canal Road to reach the boathouse.

By Metrobus   Bus routes are subject to change. Routes passing through Georgetown include 31, 32, G2, D1, D2, D3, and D6. See MetroOpensDoors.com for latest route information.

By Car   To the uninitiated, parking at the Boathouse at Fletcher's Cove can be challenging. First off, on weekdays, Canal Road, N.W., traffic flows one-way south/eastbound during the morning rush hour, then reverses to one-way westbound during the afternoon rush hour. Traffic moves fast, when it's not backed up.

It is best to arrive early to get a space. View a map or GPS beforehand if you can. Also, the entrance is easy to miss (the stone Abner Cloud House is a useful landmark) and is narrow and meets Canal Road, N.W., at a sharp angle that prevents safe right turns off Canal Road, N.W., for southbound travelers. So, only approach the parking lot headed north/west (which is not possible during weekday mornings, when Canal Road traffic only flows south/east).

From Rosslyn, Virginia, cross into the city via the Key Bridge. After crossing the bridge, turn left onto Canal Road, N.W. (not possible on weekday mornings). Keep left at the junction with Foxhall Road, N.W., remaining on Canal Road, N.W., and continue to the entrance, which is on the left side at Canal and Reservoir Roads, N.W.

By Foot   The hike from Georgetown to the Maryland line on the towpath is flat and about seven miles round trip.

To reach the canal on foot, take 29th through 33rd Streets south of their intersections with M Street. Walkways and bridges lead to the towpath.

A royal paulownia tree, or princess-tree, blooms in spring along the canal.

By Bike Bikers can take the Capital Crescent Trail—the old B&O Railway's Georgetown Branch right-of-way—from west Silver Spring in a counterclockwise half circle to the D.C. portion of the canal, arriving at the canal near the Boathouse at Fletcher's Cove (202-244-0461 or see http://fletchersboathouse.com/) and continuing along the towpath to K Street in Georgetown. The distance from Silver Spring to K Street is 11 miles. The Silver Spring to Bethesda portion is a crushed stone surface (with future plans to have it paved). The 7 miles from Bethesda to K Street are asphalt. Round-trip time between Bethesda and K Street is about one and a half hours. (For more information, see http://www.cctrail.org/.)

The Capital Crescent Trail also joins the Rock Creek Hiker-Biker Trail, providing cyclists an approximately 22-mile circular route to explore both the canal and Rock Creek Park.

NEARBY

Georgetown Reservoir Along the west side of MacArthur Boulevard, just south of Reservoir Road, Georgetown Reservoir was known for waterfowl and raptors coming and going along the Potomac River just downhill. You can view the fenced-in reservoir from the adjacent sidewalk along the west side of MacArthur Boulevard. In the past, birding was good between October and April. A typical visit might yield a few ducks, including ring-necked duck, lesser scaup, ruddy duck, bufflehead, and canvasback. Particularly during migration, large flocks appeared, increasing the likelihood of finding more species. Gulls—ring-billed, herring, and great black-backed—are sometimes joined by rarities, such as lesser black-backed gull. American coots and pied-billed grebes also show up here, as do small numbers of shorebirds. During migration, depending on weather conditions, a good number of raptors may be seen overhead. Northern rough-winged and barn swallows nest here spring to summer. Morning light is best for viewing the birds. Birding at this reservoir has not been as productive in recent years.

Great Falls and the C&O Canal in Maryland   Once a nightmare to those hoping to navigate the Potomac, the tumbling falls and roiling rapids at Great Falls, as seen from both Maryland and Virginia shores, today provide what many call the most striking natural setting near the city. In Virginia, Great Falls Park is accessible from Old Dominion Drive (fee). From Maryland, the Chesapeake and Ohio Canal National Historical Park's Great Falls Tavern Visitor Center (fee) can be reached via MacArthur Boulevard by car, and on foot or by bike via the towpath. Both parks sit about 10 miles northwest of the city. From spring to fall, parking lots fill fast on weekends and holidays.

About 23 miles northwest of the city is Riley's Lock, off Riley's Lock Road in Maryland. This site, where Seneca Creek flows into the Potomac, is of both historic and natural history interest. The aqueduct and lock here were made of red sandstone quarried only about 100 yards away. The old quarry ruins sit in the woods along the edge of the old turning basin. There you can imagine the riverboats floating where ducks, herons, swallows, and eagles now land. The red sandstone from this site was used to build the Smithsonian Castle. Local birders call this spot "Seneca," and come here to watch warblers in the trees and ducks, grebes, terns, and loons on the river.

Less than a mile south on the towpath (or reached by car via Violettes Lock Road) is another red sandstone lock called Violettes Lock. Here, the Potomac changes from calm to rippling, as it runs over gentle rapids, a change not lost on hungry water birds seeking fish and other prey.

A stroll from Riley's Lock south to Violettes Lock and back is a favorite migration hike for local birders, but other hot spots are farther south, back toward the city, including Blockhouse Point, Pennyfield Lock, and Swain's Lock. These all can be reached by car and by bike along the towpath.

Theodore Roosevelt Island

LOCATION
In the Potomac River between Rosslyn, Virginia, and the John F. Kennedy Center for the Performing Arts in Washington, D.C.

No visitor center.
Contact:
George Washington Parkway
c/o Turkey Run Park, Virginia District Ranger
McLean, VA 22101
TELEPHONE
(703) 289-2500
WEBSITE
http://www.nps.gov/this/
SIZE
88 acres
HABITATS
upland deciduous forest, floodplain forest, tidal marsh, swamp, Potomac River and shoreline

Natural History   Theodore Roosevelt Island and its smaller outlier Little Island are located in the middle of the Potomac River. The Fall Line runs down the island, with the north and west in the Piedmont and the east and south in the Coastal Plain. The Potomac River changes personality here, from rocky and rolling at the edge of the Piedmont to calmer and tidal in the Coastal Plain. This difference is easy to see on a short hike, where the rocky northern shore and then the marshy southern shore can be viewed.

Roosevelt Island is located in Washington, D.C., although its parking lot is in Virginia. A footbridge provides the only access. The river shapes the island and affects its ecology, carrying in nutrients, sweeping sediments in and out. Flood events may erode the shoreline or deposit soil in other areas. Beneath the islands is a core of mica schist, rock formed over millions of years of heat and pressure.

Trees typical of bottomland forest include pawpaw, spicebush, and American sycamore, and the park provides a rare haven for wetland plants. The island's upland forest features different plants. (See trail descriptions that follow.) In spring and fall, the island attracts many land and waterbirds following the Potomac River during migration.

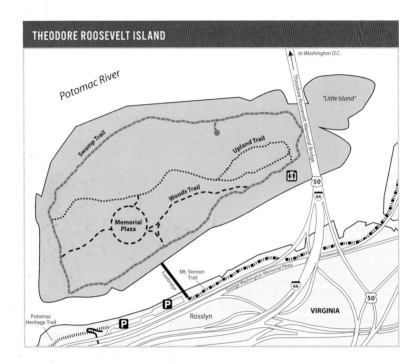

Potomac River

to Washington D.C.

Theodore Roosevelt Memorial Bridge

"Little Island"

Swamp Trail

Upland Trail

50

66

Woods Trail

Memorial
Plaza

Mt. Vernon
Trail

Footbridge

George Washington Memorial Pkwy

66

50

P

Rosslyn

VIRGINIA

Potomac
Heritage Trail

P

Human History. Walk the trails and you might think this island was always wild. But you will see few if any trees older than a century. In fact, people cleared the land and lived here, although little visual evidence remains. Long ago, a seasonal Native American fishing village occupied part of the island. A tribe living nearby was called the Nacotchtanks, or Anacostans, and in the late 1600s, the island was called Anacostian Isle and later Analostan Island. After settlement, King Charles I deeded the island to Charles Calvert, Lord Baltimore, and it was named My Lord's Island. Later, a seafarer bought the land and named it Barbadoes.

In 1717, George Mason III bought the island, which became "Mason's Island." By the late 1700s, the island was owned by his grandson, John Mason, who was a wealthy banker and businessman and the son of George Mason IV, famed writer of Virginia's Declaration of Rights and an ardent proponent of a federal Bill of Rights. Mason built a neoclassical mansion on the estate, accenting the home with extensive lawns, flower and vegetable gardens, corn and cotton fields, and trees, including apple, peach, and cherry. In 1805, a causeway was built, linking the island to Virginia's shoreline. This land link was later destroyed by a raging flood in 1877. The current

A boardwalk wends its way through the swamp at Roosevelt Island.

Visitors can spend some quality time with Teddy Roosevelt at this quiet memorial plaza within the forest.

access bridge was later built in more or less the same location. The Mason family sold the property in 1833, after which it was used for public gatherings and commercial gardens.

During the Civil War, Union forces, including African American volunteers of the 54th Massachusetts Regiment, used the island as a camp, training area, and hospital. After the war, it became an athletic club. By the late 1800s, the island was mostly grassy, open field. Few trees grew there. Fires in 1869 and the early 1900s left the mansion in ruins.

In 1932, the Theodore Roosevelt Memorial Association bought the island to create a living memorial to the 26th president, whose conservation achievements included the protection of almost 230 million acres, including 5 national parks, more than 50 wildlife refuges, and 150 national forests. The property was deeded to the federal government, and the National Park Service began managing it. Some native trees were replanted, others regrew. The mansion's remaining walls were torn down. The memorial plaza, dedicated in 1967, is now located within the island's tall secondary forest. Ongoing efforts to maintain the native vegetation include removal of invasive nonnative plants, including Japanese and bush honeysuckle and English ivy.

This island park is a fitting tribute to a vibrant president and conservation pioneer. It is an easy, wild escape from the adjacent urban areas of Washington, D.C., and Rosslyn, Virginia. You can immerse yourself in the woods and marsh, even if the whirr of parkway and bridge traffic and the roar of airliners headed to and from nearby Ronald Reagan Washington National Airport remind you that you're in a busy city. Yet many area residents never visit this urban refuge, barely noticing the woodland that blurs by on their commutes up and down the George Washington Memorial Parkway.

The Memorial Plaza   Ringed by rows of tall willow oaks and surrounded by the forest, the Memorial Plaza features a 17-foot-tall sculpture of Theodore Roosevelt flanked by 21-foot-tall granite tablets highlighting his philosophical tenets. This amphitheater-like setting is dotted with benches and is a nice, contemplative spot. But it can also be fun for the restless birder during spring and fall migrations, when warblers, tanagers, grosbeaks, thrushes, flycatchers, and other land birds flit through the canopy and undergrowth, particularly early in the morning. Late fall to early spring, hermit thrush and winter wren may show up here.

The Shoreline   In the woods you feel far from the hustle and bustle, but from the shoreline you can see you are in the thick of things. Georgetown is to the north, the John F. Kennedy Center for the Performing Arts to the east, and Rosslyn, Virginia, to the west. Ring-billed gulls are a familiar sight, joined by some migrating Caspian terns in spring, late summer, and early fall. Also, watch for up to six swallow species, including locally nesting northern rough-winged swallows and barn swallows, which nest under the entrance bridge. Osprey (spring and summer), bald eagles, and particularly double-crested cormorants often flap up or down the Potomac, and nearshore, northern water snakes sometimes bask in the open or swim past. The Swamp Trail, footbridge, and parking lot on the Virginia side offer nice river views.

The tidal flats just south of the island can be exciting for water-birds. From the parking lot, walk the Mount Vernon Trail (watching for cyclists as well as birds) south until you see the flats, which are most productive at low tide.

The introduced aquatic plant hydrilla grows in clumps that show at the water's surface. This plant attracts waterfowl and at low-tide shorebirds may forage in and around it.

The Swamp Trail   Of the island's three main trails, this is the longest (1.3 miles) and most important from a wildlife perspective. Upon crossing the footbridge onto the island, the Swamp Trail can be walked either left or right. It circles the island, affording a few vantage points of the Potomac shoreline.

Elevated boardwalk with a few sitting areas makes up about a third of the Swamp Trail; the rest is a wide, level dirt path. A short turnoff to a marsh overlook is a highlight. Water hemlock, New York ironweed, pickerelweed, yellow flag iris (also called pale-yellow iris), cattail, green arrow arum, black willow, and baldcypress trees (undoubtedly planted at some time) highlight this viewpoint. From spring through fall, watch for wood duck, green heron, common yellowthroat, indigo bunting, and red-winged blackbird. In summer, pickerelweed and other wetland flowers attract tiger and spice-bush swallowtails, while dragonflies and damselflies hunt tiny flying insects just above the water's surface.

Part of the trail traverses wooded swamp presided over by massive silver maples. The swamp usually harbors woodpeckers and, during migration, waterthrushes and other warblers, including, with luck, the prothonotary warbler. In winter, watch for white-throated, song, swamp, and perhaps other sparrows, brown creepers, and winter wrens. In spring and summer, eastern kingbirds, Baltimore orioles, and warbling vireos grace the treetops as they feed, sing, and protect their breeding territories.

As the trail loops around the southern end of the island, it passes over a tidal gut and channel between Theodore Roosevelt Island and its tiny sister, Little Island. Low tide reveals mud that may draw a killdeer or other shorebird, when they are present. Great blue herons, wood ducks, and mallards often loaf or feed here. In early morning or at dusk, beavers may be seen. Raccoons' small hand-like prints may decorate the mud.

Upland and Woods Trails   These shorter trails highlight the higher, drier woodland, which includes such tree species as northern red oak, ashes, and, here and there, American beech and American holly. Somewhere off the trails, at a high point in these hilly woods, sit the buried remains of Mason's mansion.

The Upland Trail runs through the middle of the island and is 0.7 miles long. The Woods Trail starts near the footbridge entrance and runs into the Swamp Trail. The trails are wide, but visitors should be cautious along the edges, watching for poison ivy, stinging nettles, and ticks. In sunny patches in the forest, watch logs and stumps for a glimpse at the five-lined skink, or perhaps a black rat snake. An eastern box turtle may walk or sit among the leaf litter. The only likely clue to gray treefrogs' presence will be their slow but loud trills. Easier to see will be five woodpecker species, from the tiny downy woodpecker to the near crow-sized pileated, plus the more retiring yellow-bellied sapsucker, although this species is only present late fall to early spring. Watch for active and vocal flocks of Carolina chickadees and tufted titmice, which often travel in company of kinglets, nuthatches, creepers, and migrating warblers. Red-shouldered hawks and barred owls frequent these woods and sometimes nest on the island. Cooper's, sharp-shinned, and red-tailed hawks also regularly appear. If searching for barred owl, check large tree cavities and watch for agitated groups of small birds mobbing this large, streaked and barred raptor. Fall foliage includes a mixed palette of oak, ash, hickory, maple, and dogwood leaves. You are likely to see gray squirrels and white-tailed deer, part of a growing population of these ungulates that has dramatically thinned the forest undergrowth in recent years.

GETTING THERE

By Metrorail   The Metro's Rosslyn station (on the Blue and Orange Lines) is the nearest to the park. See the detailed map at the Rosslyn station to find directions for the short walk, in the direction of Key Bridge, to the pedestrian/biker overpass over the George Washington Parkway. The overpass leaves visitors on the east side of the parkway, at the north end of the Roosevelt Island parking lot (and the Potomac Heritage Trail and Mount Vernon Trail). This walk takes 10 to 15 minutes.

By Car   Theodore Roosevelt Island is accessible only from the northbound lanes of the George Washington Memorial Parkway.

All visitors enter and exit Roosevelt Island via this bridge from Virginia.

The entrance to the parking lot is located just north of the Roosevelt Bridge (Route 66).

If coming southbound on the George Washington Parkway, take the Theodore Roosevelt Bridge into Washington, to Constitution Avenue, N.W. Once on Constitution, take a left onto 22nd Street, N.W., and then a quick left onto C Street, N.W., followed by another left onto 23rd Street, N.W. You will then turn right, completing a square that takes you back to the Theodore Roosevelt Memorial Bridge, but westbound this time. Across the river, bear right to return to the George Washington Memorial Parkway. The Roosevelt Island parking lot will soon appear on the right.

If driving from inside Washington, cross the Arlington Memorial Bridge toward Virginia, bearing right to go to the George Washington Parkway. The parking lot appears on the right, just after passing the Theodore Roosevelt Bridge.

You must leave the park via the George Washington Parkway's northbound lanes. To return to southbound lanes, after about a mile take the left-hand exit for Spout Run. About a half mile after this turnoff, take the left lane to a U-turn that takes you south on Spout Run, leading back to George Washington Parkway's southbound lanes.

By Foot or by Bicycle From Rosslyn, Virginia, which overlooks the Potomac and the island, walk over the bicycle and footbridge, which overpasses the George Washington Parkway and leaves off near the Roosevelt Island parking lot and entrance. To reach this from Rosslyn, you can take Lynn Street over Interstate 66 and across Lee Highway and then turn right onto the paved Custis Trail, which leads downhill to the overpass.

The park is easily reached by bike, off the Potomac Heritage Trail (from the north) and Mount Vernon Trail (from the south). These trails are one in the same but change names at the parking lot. They parallel the busy George Washington Parkway and are one of the most popular and scenic bike rides in the area. Bike racks are located near the footbridge that leads from the parking lot to the island.

Dumbarton Oaks Park, Montrose Park, and Dumbarton Oaks and Gardens

LOCATION

Three very different, adjacent parks in Georgetown.

Two public parks administered by the National Park Service—Dumbarton Oaks Park and Montrose Park—are reached via Lovers Lane (foot traffic only), which runs north and downhill from R Street, N.W., between 30th and 31st Streets, N.W.

Dumbarton Oaks and Gardens, about 10 acres of formal gardens, is a private property with afternoon hours and an admission fee (from mid-March to the end of October). Admission is free November to early March, and the property is closed on federal holidays and during inclement weather. The gardens entrance is at R and 31st Streets, N.W. The museum (free), museum shop, and music room entrance is at 1703 32nd Street, N.W.

TELEPHONE

For Dumbarton Oaks Park and Montrose Park, contact the Rock Creek Park Nature Center at (202) 895-6070.

For Dumbarton Oaks and Gardens, contact (202) 339-6401.

WEBSITES

http://www.nps.gov/nr/travel/wash/dc11.htm

http://www.doaks.org

SIZE

Dumbarton Oaks Park, 27 acres; Montrose Park, 16 acres; formal gardens of Dumbarton Oaks, 10 acres of formal gardens (total property owned by Harvard University, 16 acres).

HABITATS

formal gardens, forest, meadows, rolling lawns, stream

Natural History   Georgetown is known for its trendy shops, top-priced real estate, and history as a port town that predates the rest of Washington, D.C. But the area has natural assets as well. Today, "wild escapes" in this part of the city mean peaceful strolls through second-growth woods or along garden pathways shaded by aged ornamental trees. While not exactly the forest primeval, these are pretty parks where you can surround yourself with interesting plant life, butterflies, birds, and squirrels.

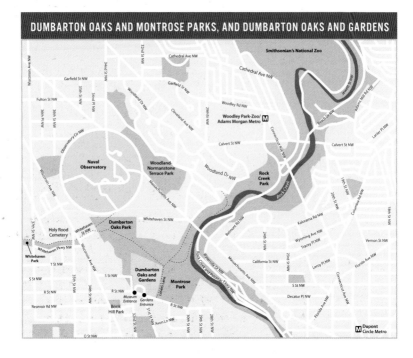

Human History The three parks in this account form a varied green space between immense and winding Rock Creek Park to the east and Glover-Archbold and Whitehaven Parks to the west.

In 1702, Scottish immigrant Ninian Beall bought a large tract of land in what is now Georgetown and named it Rock of Dumbarton, after the town where he was born, Dumbarton on the Clyde. Large eastern white oaks still accent these gardens and parks. The trees and the Beall's property designation inspired the present-day names Dumbarton Oaks Park and Dumbarton Oaks and Gardens.

While the private Dumbarton Oaks and Gardens is a well-known Washington landmark, the two public parks fall into the "hidden gems" category, known best to local residents.

Dumbarton Oaks Park Part tame and part wild, Dumbarton Oaks Park is a mix of forest and maintained and overgrown gardens that encompasses former parts of the adjacent estate. Here you will find native trees, including American beech, tuliptree, flowering dogwood, and huge eastern white oaks for which the property is named. Some parts of the park are a tangle of vines and shrubs. Others are cloaked in an undergrowth of native shrubs, including spicebush. Still other areas contain old garden plantings, including

Stroll through these parks and enjoy a hilly landscape that blends well-manicured gardens and flowering ornamentals with adjacent rustic woodland.

rhododendrons, forsythia, and a wide variety of bulbs. Meadow areas are also interspersed through the park.

At midmorning, watch for birds bathing in the stream. The tangles and trees often hold surprises during migration. Depending on the season, these might include winter wren and fox sparrow or thrushes, warblers, vireos, and orioles.

Mr. and Mrs. Robert Woods Bliss hired landscape architect Beatrix Ferrand to design a landscape that evoked the countryside below the formal gardens at Dumbarton Oaks. The couple later deeded this land as a public park. Today, the National Park Service manages it and Montrose Park, both as green spaces and cultural landscapes, former country estates that are now public parks.

Montrose Park Montrose Park has gardens, tennis courts, playgrounds, sprawling green lawns, and huge trees, including eastern white oaks, tuliptrees (including the city's largest-known specimen, which is also the city's second-largest tree), elms, American beech, and conifers. This is a popular picnic spot and the hills have been used in winter for sledding and sunbathing in summer. Lovers Lane runs along the park's western boundary. Large trees fringing the tennis court area often attract migrating birds. Check the gardens for a varying selection of butterflies, depending on season and weather. These may include the mourning cloak, which may be out on warm days in late winter.

In the early nineteenth century, the land belonged to an industrialist named Robert Parrott. Back then, Parrott allowed locals to picnic and stroll on his land, part of which was used to manufacture

Forest pathways in Dumbarton Oaks Park and Montrose Park attract migrant songbird flocks in spring and fall.

rope for his business, the Georgetown Wool & Cotton Factory. (Remnants of the ropewalk can be found in the park.) Parrott died in 1822 and the property had several owners until 1911. That year, Congress approved $110,000 payment to buy the land. Thus, it became Georgetown's first public park. In the park, a memorial to Sarah Louise "Loulie" Rittenhouse commemorates the woman who spearheaded efforts to save this important green space and make it a public place.

Dumbarton Oaks and Gardens This famous site is well known not only for its 10 spectacular acres of formal gardens but also for its Byzantine and pre-Columbian collections and libraries, which are housed in a federal-style mansion built in 1801. (The museum and library, which focuses on Byzantine, pre-Columbian, and landscape studies, are open to the public and called the Dumbarton Oaks Research Library and Collection.) In 1920, Robert Woods Bliss and Mildred Barnes Bliss bought the house and property, which was then called The Oaks. They renamed it Dumbarton Oaks. The gardens were designed and developed over almost 30 years, with the assistance of famed landscape architect Beatrix Ferrand. Accented with stone paths, walls, fountains, pools, sculptures, and iron gates, they embrace both European and distinctly American styles. In 1940, the Bliss family gave the mansion and gardens to Harvard University.

The property has a rich history. During his tenure as secretary of war (1817–1825), congressman and later vice president John C.

Dumbarton Oaks and Gardens.

Calhoun lived here. At the time, the property was called Oakly. Later, the property took on the old name for the sector of town that became Georgetown in 1751. Igor Stravinsky's Concerto in E-flat, also called the *Dumbarton Oaks Concerto*, premiered on the property in May 1938. Bliss had commissioned the composer to create the piece for the couple's 30th wedding anniversary. In 1944, diplomatic meetings at the mansion paved the way for the formation of the United Nations.

Dumbarton Oaks mansion is located at the highest point in Georgetown. Because of the steep slope descending below the house, the gardens were terraced to blend with the natural contour of the site. From the mansion, you can look down at the gardens and see the Rock Creek Valley below.

The property only opens in afternoons from 2 p.m. to 6 p.m. from March 15 to October 31, and to 5 p.m. November to March 14. This limits birding opportunities, but on warm afternoons from May to early fall, butterflies and dragonflies are active. The mixture of elegant landscaping; a diversity of trees, shrubs, and flowers; and the site's overall serenity make the gardens worth a visit at any season.

Spring and fall, however, are the most popular times, when weather is ideal and many plants are either flowering or sporting bright fall foliage. Early spring highlights include simultaneous peak blooms on Cherry and Forsythia Hills. A bit later, Crabapple Hill comes into its glory. But there are blooms for almost all seasons,

from early spring daffodils, tulips, and crocus to fall chrysanthemums. March to May and September to October, expect lines on
weekends, especially at opening time at 2 p.m. Dumbarton Oaks is
closed on Mondays, federal holidays, and during inclement weather.
Picnicking is not permitted.

GETTING THERE

By Foot These parks can be reached off R Street, N.W., between 30th and 31st Streets, N.W., via Lovers Lane. Lovers Lane runs
along the western boundary of Montrose Park and the eastern
boundary of Dumbarton Oaks and Gardens. (The entrance to Dumbarton Oaks and Gardens, however, is at R and 31st Streets, N.W.) At
the bottom of the hill, the Dumbarton Oaks Park entrance appears
on the left, while on the right, a trail follows a stream east through
Montrose Park to a nearby bend of Rock Creek.

You can create longer hikes, reaching Dumbarton Oaks and
Montrose Parks via footpaths from adjacent Rock Creek and Potomac
Parkway, to the east, or via a hiking route from the west, from
Glover-Archbold Park to Whitehaven Park, crossing Wisconsin Ave-
nue, N.W., to a dead end at Whitehaven Street, N.W., at the north-
west corner of Dumbarton Oaks Park. From here, you can turn onto
this park's path (described below), or you can continue to a park at
Normanstone Parkway, where there is a link to Rock Creek and
Potomac Parkway as well.

The trail through the interior of Dumbarton Oaks Park continues east, reaching Lovers Lane, then Montrose Park, then, a bit to
the east, the Rock Creek and Potomac Parkway. (See map.)

By Metrorail The Dupont Circle (Red Line) station is a mile
walk to the parks following these directions: From Dupont Circle,
take Massachusetts Avenue, N.W., in a northwest direction. Turn left
onto Q Street, N.W. (crossing Dumbarton Bridge), and walk several
blocks. Turn right onto 31st Street, N.W., and walk up the hill to
reach R Street, N.W. (See park entrance locations above.)

The following bus routes pick up near the Dupont Circle station
and pass near the parks: D-1, D-2, D-3, and D-6.

The Woodley Park-Zoo / Adams Morgan (Red Line) station is
another possibility, closer as a crow flies but not as direct a walk as
the Dupont Circle station.

By Metrobus Metrobuses 31, 32, 36, D-1, D-2, D-3, and D-6
come within a few blocks of the entrance areas. See MetroOpens-
Doors.com for latest bus route information.

By Car   There is two-hour street parking near the entrances, but heed the time limit to avoid being ticketed on all but Sunday, when there are no time limits. An alternative is parking about a 15-minute walk away, in public lots off Wisconsin Avenue, N.W.

To reach Lovers Lane (Dumbarton Oaks and Montrose Parks), head north on Wisconsin Avenue, N.W., and then turn right on R Street, N.W., and park near 31st Street, N.W. For pedestrians only, Lovers Lane is just east of 31st Street, N.W.

NORTHEAST

U.S. National Arboretum

LOCATION
About a 10-minute drive (2 miles) east of the U.S. Capitol, with two entrances: one at 3501 New York Avenue, N.E. (Route 50), and another off Bladensburg Road, off 24th and R Streets, N.E.

The grounds are open 8 a.m. to 5 p.m. Friday through Monday, except December 25.

TELEPHONE
(202) 245-2726

WEBSITE
http://www.usna.usda.gov

SIZE
446 acres, with 9 miles of paved, winding road and many trails

HABITATS
ornamental gardens, meadows, conifer groves, woodland, mixed gardens with canopy of native trees, river's edge, ponds, and streams

Natural History Loblolly pine, a southern tree representative of the Coastal Plain, grows within the mixed forest here, along with the shorter-needled Virginia pine. Pine warbler nests here, and this is one of only a few places it can be easily found within the city. Other songbirds nesting in the arboretum's forest areas include great crested and Acadian flycatchers, red-eyed vireo, blue-gray gnatcatcher, wood thrush, and northern parula warbler.

The pines also may attract wintering red-breasted nuthatches, their numbers varying depending on the year. In any wooded areas, watch for pileated, red-bellied, downy, and hairy woodpeckers and northern flickers and yellow-bellied sapsucker from fall to spring.

Because of its large size and varied habitats, the arboretum provides an important refuge for amphibians and reptiles. Green frog, spring peeper, and southern leopard frog are common, and in early spring, spotted salamanders may be seen. Reptiles include black rat snake, eastern garter snake, eastern painted turtle, and the red-eared slider.

Raccoons are sometimes seen during early morning forays. Woodchuck, eastern cottontail, and red and gray foxes also live here.

Human History Perched on the shores of the Anacostia, the National Arboretum property likely attracted bands of hunting and

U.S. Capitol columns and adjacent meadow create an almost Grecian setting rich in butterflies, wildflowers, and other wildlife.

gathering Paleoindians. The area was certainly visited later on by Algonquian-speaking Native Americans, who had their main village nearby. More than 30 potential archaeological sites have been identified on the property, but only two have been fully investigated and cataloged.

In the late 1600s, the location where the arboretum now sits likely belonged to one or two English landowners. In the arboretum area, land grants were held by Colonel Ninian Beall, one of the first representatives of Prince George's County in the 1696 Maryland General Assembly, and a sailor named Benjamin Hadduck.

When Washington became the capital city, the arboretum fell outside the original plan area. By the 1800s, small farmsteads and orchards dotted the land, along with thin ribbons of woodland. The landscape remained rural into the 1920s.

In 1925, in the journal *Science*, Frederick V. Coville, a U.S. Department of Agriculture botanist soon to become the arboretum's first director, wrote: "The proposed national arboretum at Washington would contain a permanent living collection of trees and other outdoor plants for purposes of scientific research and education. It would include the trees, shrubs, and perennials used in forestry and horticulture, and the wild relatives of these plants. It would be a bureau of standards for horticulture . . . and it would serve incidentally as a bird sanctuary."

In 1927, the U.S. government approved the land purchase and establishment of the only federally funded arboretum. This included buying the Mount Hamilton and Hickey Hill tracts from about 30

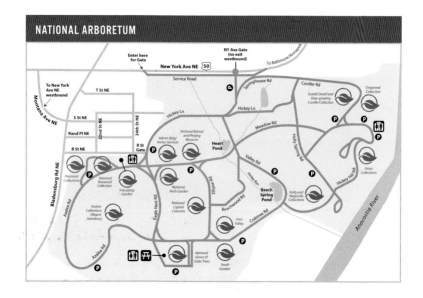

owners. The 1925 estimated value of these 408 acres was $343,000. The facility did not fully open to the public until 1956.

Today, the arboretum brochures call it "a U.S.D.A. research and education facility and a living museum." New plant breeding techniques and new cultivated varieties (cultivars) are developed here, mostly floral and landscape plants. Over the years, the arboretum introduced more than 670 cultivated varieties, or cultivars. The facility is also a storehouse of plant information.

This expansive green space could also be called the capital's "secret garden" because although it sits in plain sight, many residents and visitors have never visited. Visitors stroll extensive gardens where many of the plants are labeled. About half of the property is gardens; the other half (about 155 acres) is forest and meadow.

A visitor will see different blooms and wildlife on each visit, depending on luck, weather, and time of year. Check at the Administration Building near the entrance for information on tram tours, exhibits, lectures, and hikes. Or see the website above. Admission and parking are free.

A good way to see the arboretum is to bring your bike and ride the roads. For short walks, options include four adjacent areas: the aquatic plants and koi pool at the Administration Building, the Friendship Garden, the National Herb Garden, and the National

One of the largest azalea collections anywhere provides
a feast for the eyes from late April into May.

Bonsai & Penjing Museum (one of the country's largest bonsai collections).

Featured Blooms and Seasons   Something is always on show at the arboretum. Here are some highlights:

January to early March: witch-hazels in subtle bloom; hollies and conifers stand out.

Late March to early April: early magnolias and flowering cherries, spring bulbs (including daffodils and crocuses) in bloom.

Mid-April: flowering cherries and woodland wildflowers add bursts of color.

Late April into May: azaleas on Mount Hamilton (until about the third week of May), rhododendrons, crabapples, dogwoods, magnolias, peonies, and roses in bloom. Peak of spring songbird migration.

June, July, and August: mountain laurels (late May to mid-June), crape-myrtles, water lilies at the Administration Building, day lilies, and herbs in bloom. Meadow and prairie flowers, including butterfly weed, in bloom. Watch for butterflies, dragonflies, and damselflies.

September to October: showy crabapple, viburnum, and firethorn fruits attract birds. Fall foliage reaches peak. Meadow wildflowers bloom. Chrysanthemums (October to November). Peak of fall songbird migration. Hawk migration in October.

November to December: holly berries attract robins, cedar waxwings, and other songbirds. Chrysanthemums. Hawk migration in November and by December leafless trees make watching overwintering raptors and other birds easier.

Meadow and woodland edge habitats shelter eastern cottontails, which in turn provide food for resident foxes and raptors.

Exploring the entire U.S. National Arboretum would take a few visits. Exhibits include the National Bonsai & Penjing Museum, the Holly and Magnolia Collections, the National Boxwood Collection, the National Herb Garden (800 kinds of herbs), perennial collections (daffodils, peonies, day lilies), and a youth garden. The arboretum houses an extensive herbarium with more than 650,000 dried, pressed plant specimens and a botanical library with more than 11,000 books.

These areas provide some of the best wildlife-viewing opportunities:

Capitol Columns Meadow Twenty-two Corinthian columns adorn this scenic wild meadow, which is maintained both to attract wildlife and to cut mowing costs. Carved from Virginia sandstone in the 1820s, the columns were removed from the U.S. Capitol's East Portico during its 1958 expansion. Reminiscent of a Grecian landscape, this grassy hill is flanked by a reflecting pool and wildflowers, including asters, milkweed, and goldenrod, which attract monarchs, swallowtails, pearl crescent, common buckeye, clouded sulphur, and many other butterflies.

Dogbane, native grasses, and other meadow plants flourish here. A wide mowed path leads through the meadow. In spring and early summer, listen and look for American goldfinches, eastern bluebirds, indigo buntings, blue grosbeaks, eastern towhees, and chipping sparrows. The buntings may sing from scattered pin oaks, where eastern kingbirds often perch from spring to early fall. Sparrow flocks frequent the area in fall and winter. Red-tailed hawks hunt in the area. Watch for black rat snakes and eastern garter snakes here and elsewhere on the property. This and other open areas also host eastern cottontail, woodchuck (also called ground-hog), and red fox, best seen early in the day. The elusive gray fox lives on the property but is rarely seen.

Azalea Collections More than 15,000 azaleas grow on Mount Hamilton (elevation 240 feet) beneath such native trees as flowering dogwood, eastern white oak, and black tupelo (black gum). From late April into May, this is the best azalea show in town. The woodland setting and dense undergrowth attract eastern towhee, brown thrasher, migrating thrushes, sparrows, warblers, tanagers, and vireos. From the hilltop, you get a great view of the U.S. Capitol. This can be a good hawk-watching spot as well.

Fern Valley A well-marked half-mile trail rambles through woodland and meadow at the arboretum's collection of featured eastern U.S. plants. Here, Washington, D.C.'s native trees shade plants grouped to represent the following eastern habitats: northern forest, Piedmont region, southern mountains, southern lowlands, and southern Coastal Plain. Walking the trail is like visiting a living field guide: Many of the wildflowers, ferns, trees, and shrubs are labeled. The trail crosses a gurgling stream, which leads to a small pond. For the naturalist interested in local plants, butterflies, and birds, Fern Valley easily warrants a slow morning's walk and exploration.

Beech Spring Pond During warm days in late winter, listen here for the first spring peeper choruses. The pond attracts Canada geese, ducks, and a few herons. During summer, annual cicadas buzz from the tall trees here and elsewhere on the property. Also check the smaller Heart Pond, located a bit further down Valley Road from this pond. Red-eared sliders and eastern painted turtles bask on logs and along the water's edge, and snapping turtles cruise the murky shallows. Beavers have built dens on the property at Beech Spring Pond, Spring House Pond, and the Hickey Run stream.

In warm months, Beech Spring Pond and other wet areas of the arboretum can be very productive for dragonflies, including common green darner, common whitetail, eastern amberwing, and damselflies, including violet dancer, orange bluet, and stream bluet.

Asian Collections These encompass 13 acres of Asian perennials, vines, shrubs, and trees, including many close relatives to such North American natives as jack-in-the-pulpit and tuliptree. Exhibit areas here are Asian Valley, China Valley, Japanese Woodland, Korean Hillside, and the Camellia Collection. Some of the arboretum's oldest plantings grow in the Asian Valley section, dating back to the 1940s. This part of the arboretum overlooks the Anacostia River. Arboretum expeditions to China, Japan, and Korea helped bring once-rare but now familiar ornamentals to U.S. gardens, including chrysanthemums, hostas, pachysandra, cold-hardy camellias (many cultivars developed at the arboretum), and bamboo, all of which flourish in this part of the arboretum.

Dogwood Collection   Likely the largest collection of dogwoods in the region, this area also affords a nice view of the Anacostia River.

Gotelli Dwarf and Slow-Growing Conifer Collection   The National Arboretum contains some of the largest conifer groves in the city. Many consider this one of the best dwarf conifer collections in the world. Fall to spring, check these five acres for songbirds, including golden-crowned kinglets and red-breasted nuthatches. A slow search for roosting owls might bear fruit. The next two sites also have large groves of conifers.

National Grove of State Trees   Hailed as a "living monument to America's forest resources," here you will find a representative tree for each state. Warblers, kinglets, nuthatches, flycatchers, and sometimes owls, are among the birds seen here. The arboretum's only picnic area is located at this site.

Hickey Hill and the Anacostia River   Hickey Hill is a forested rise covered in deciduous forest peppered here and there with redcedar, American holly, and large white pine and spruce. Birders scour these woods for migrant songbirds, hawks, and other avian surprises. The Anacostia River lies to the east. From its shores, visitors may see ducks, bald eagles, and osprey. Two access points allow arboretum visitors to reach the Anacostia River shore: from the bottom of China Valley and via a dirt road starting adjacent to the Flowering Tree Collection at the western end of Hickey Hill Road. Gates to these Anacostia access points are open 8:30 a.m. to 4 p.m.

Friendship Garden   This living demonstration landscape surrounds the Arbor House, which contains the arboretum's gift shop, some offices, and restrooms. The plantings of perennials, shrubs, and trees are not only eye catching but designed to reduce water use and maintenance costs. They also attract wildlife, including butterflies and birds, such as eastern towhees and brown thrashers.

April is the best time for tulip watching at the
National Arboretum and in many other D.C. parks.

GETTING THERE

By Car From the Capital Beltway (I-495/I-95) in Maryland, take exit 22B, the Baltimore–Washington Parkway, west toward Washington. After about 7 miles, take Route 50 (New York Avenue) west. Get in the left lane and take a left onto Bladensburg Road. In four blocks, take a left on R Street, N.E. In two blocks, you will see the National Arboretum's gate.

From inside Washington, from the Northwest quadrant, head east on New York Avenue to the intersection with Bladensburg Road, N.E. Turn right (south) onto Bladensburg Road, N.E., go four blocks to R Street, N.E., and make a left. Continue two blocks to the gate.

If coming from Virginia, the National Arboretum can be reached via several major roads, including the Capital Beltway (see above) via 95 North, I-95 to I-295, I-395, and I-66.

By Metro The Blue and Orange Lines' Stadium Armory Station is the closest Metrorail station. From there, transfer to the B-2 Metrobus that heads north. Soon after leaving the station, the bus runs up Bladensburg Road, N.E., where you should exit at the intersection with R Street, N.E., and walk two blocks east on R Street to the entrance. See MetroOpensDoors.com for up-to-date bus information and a map showing this route.

By Bike The hiker–biker Anacostia Riverwalk Trail will link the National Arboretum on the Anacostia River's west bank with a portion of Anacostia Park on the east bank, including Kenilworth Aquatic Gardens (see separate entries for both of these sites).

U.S. NATIONAL ARBORETUM

95

Kenilworth Aquatic Gardens

LOCATION
Kenilworth Aquatic Gardens, across the Anacostia River from the National Arboretum, is managed by the National Park Service under Anacostia Park. (See separate entry.) It is treated here as a separate entry because of its distinct history, wildlife, habitats, and facilities.

TELEPHONE
(202) 426-6905

WEBSITE
http://www.nps.gov/keaq

SIZE
12 acres of aquatic gardens and more than 70 acres of wetland, administered as part of the 1,200-acre Anacostia Park

VISITOR CENTER
The visitor center is open daily from 8 a.m. to 3:45 p.m. The park has a parking lot, restrooms, and an intermittently staffed bookstore. The gardens are open from 7 a.m. to 4 p.m. and are closed only on Thanksgiving, December 25, and January 1.

HABITATS
freshwater tidal marsh, ponds, eastern deciduous woodland typical of that found in the Coastal Plain

Natural History   Kenilworth Aquatic Gardens provides the visitor with a rare chance to study wildlife found few other places in the city. The artificial ponds harbor wetland wildlife, as does the stunning, revitalized marsh, a remnant of habitat once found along much of the Anacostia River.

Human History   In the 1880s, Civil War veteran Walter B. Shaw bought 32 acres along the Anacostia River. He built dikes in the wetland adjacent to his farmland, creating ponds where he planted some water lilies, sparking a hobby and business that took off around 1908 and would continue for many years. His daughter Helen Fowler traveled with him to collect exotic lilies. They propagated them and developed new varieties. Shaw died in 1921, but Fowler continued to run the business. From 1921 to 1938, the gardens were a local, commercial attraction. Thousands of visitors strolled the gardens and picnicked and partied there.

In the 1930s, the Anacostia River adjacent to the gardens was dredged and most of the wetlands were filled in. As Army Corps of Engineers dredging operations moved upriver, the ponds' future was threatened. Fowler fought for years to protect her ponds from

Kenilworth Aquatic Gardens' boardwalk provides rare access to one of the city's last tidal marshes.

being filled. In 1938, Congress allocated money to buy the eight acres of water gardens for $15,000 and add it to Anacostia Park. As part of the deal, Fowler lived on the property until she died in 1957. Once called the W. B. Shaw Lily Ponds and Shaw Gardens, the property was renamed after the local community by the National Park Service.

In 1993, 33 acres of marsh were restored in the park, augmenting the park's wetland area to about 70 acres. This cooperative project enlisted federal, city, and private groups seeking to improve water quality and wildlife habitat. Today, this habitat is thriving and can be viewed from the end of the Boardwalk and River trails. This marsh is one of the city's only accessible wetlands that recall the extensive marshes of yesteryear.

The Ponds Forty-five ponds dot the property. They are filled with hardy water lilies, lotuses, and tropical water lilies that can only be put out after the threat of frost. To see these aquatic plants in bloom, visit between late May and early September. July is the peak. During this month, an annual Lotus and Waterlily Festival is held at the park. In August, 6-foot-wide Victoria lilies might be on show.

On an early morning walk, you may see a beaver in the park's wetlands. There is an active beaver lodge, and you will likely see gnaw marks and toppled trees, evidence of the industrious rodents. Great egrets, great blue herons, green herons, and sometimes night-herons can be seen close up, hunting the ponds' crayfish, fish, and amphibians. In the ponds and adjacent wetland live the following:

Great egret is one of several heron species found in the park.

bullfrogs, green frogs, American and Fowler's toads, spring peepers, and pickerel and southern leopard frogs.

On warm days, red-eared sliders and eastern painted and redbelly turtles bask at the water's edge. Sometimes eastern mud and spotted turtles show up as well. In late spring and summer, dragonflies abound. Kenilworth Aquatic Gardens is the only place in the city where unicorn clubtail is regularly seen. Baldcypress grow along the edges of some ponds, the trees' "knees"—exposed portions of the roots that protrude above the soil—dot the shore. These trees were planted here. While they don't grow wild in D.C., the city is not far from the northern edge of this tree's native range (Calvert County, Maryland, and southern Delaware). Common wildflowers include swamp milkweed (host plant to the monarch butterfly), joe-pye-weed, ironweed, jewelweed, and purple loose-strife, a nonnative invasive plant the park staff work hard to control.

The River Trail   This 1.4-mile (0.7 miles each way) trail leaves the ponds behind, passing through forest, skirting the marsh, and ending by the river. Among the trees growing here are sweetgum, pawpaw, pin oak, and willow oak. During spring and fall migration, this and the boardwalk trail are fine places to watch warblers, swallows, spotted sandpipers, and other migrants. Prothonotary warbler may be singing here spring to summer, and watch for indigo buntings and orioles as well. Wood ducks nest and red-shouldered hawks often call and fly through the woods.

At the end of both trails, osprey and bald eagle are regularly seen. In addition to turtles, reptiles often seen along this trail include black rat snake, ribbon snake, and northern water snake, and the

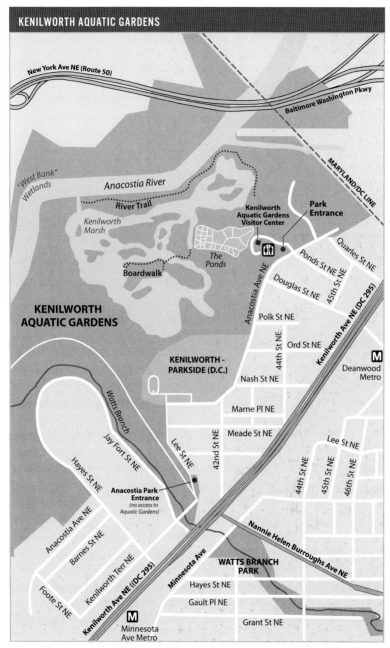

five-lined skink, a sleek, striped lizard frequenting rotten stumps and tree falls. Zebra swallowtail, found here close to the northern limit of its range, uses pawpaw as a host plant. Watch also for spicebush swallowtail (uses spicebush as a host plant) and hummingbird or sphinx moths.

Anacostia River fish include the introduced snakehead, largemouth bass, perch, sunfish, and catfish.

Flowering lotus.

The Boardwalk   The boardwalk is pretty much the only place in the city where you can immerse yourself in a marsh without getting your feet wet. The restored marsh attracts a wide variety of wildlife. Male red-winged blackbirds sing and show off their red epaulets, orchard orioles hold territory late April well into the summer, and from early spring through summer, tree swallows fly past, the sun glinting off their metallic blue-green backs.

October and November are peak months for sparrow migration, but many linger through winter. Muskrats live in the marsh and can often be seen along the boardwalk trail. They resemble small beavers with rat-like tails. Pickerelweed, crimson-eyed swamp mallow (a native hibiscus), cattail, and some patches of native blue flag iris and wild rice grow in the wetland. The yellow flag iris abounds. Park staff are trying to control this introduced plant.

The Visitor Center   The clover-filled fields behind the visitor center attract eastern cottontail rabbits. A few interesting trees grow there, including a huge southern magnolia, a Himalayan spruce, and a rare Franklinia tree, which is in the tea family.

GETTING THERE

By Metrorail   From the Deanwood Metro station on the Orange Line, it is about a 20-minute walk to the park entrance. From the Deanwood Metro station, go downstairs, take the Polk Street exit to the left, and walk west to the pedestrian overpass over Kenilworth Avenue (DC 295). Once over Kenilworth Avenue, walk west on Douglas Street, N.E. At the end, turn right on Anacostia Avenue, N.E. The park is about a half block on the left.

By Metrobus   See MetroOpensDoors.com for current routes.

By Car   From the Capital Beltway (I-495) in Maryland, exit onto the Baltimore Washington Parkway, southbound to the city. Shortly after it merges with Kenilworth Avenue (DC 295) south and crosses New York Avenue (Route 50), stay to the right. Exit at signs for "Eastern Avenue and Aquatic Gardens." Follow the service road parallel to 295 and take the second right onto Douglas Street, N.E. At the end of Douglas Street, N.E., make a right onto Anacostia Avenue, N.E. The park is on the left. Park in the lot. From there, a path enters the park.

In the city, from New York Avenue (Route 50), take Kenilworth Avenue (DC 295) south and stay in the right lane for the Eastern Avenue exit. At Douglas Street, N.E., turn right. At the end of Douglas Street, N.E., turn right onto Anacostia Avenue, N.E. The park entrance appears soon after on the left, at 1550 Anacostia Avenue, N.E. (between Douglas and Ponds Streets, N.E.).

From the south, take DC 295 north to Eastern Avenue and make a U-turn to the left onto the southbound service road and then turn right onto Douglas Street, N.E. At the end of Douglas Street, N.E., turn right onto Anacostia Avenue, N.E. Soon after, the park entrance appears on the left, at 1550 Anacostia Avenue, N.E. (between Douglas and Ponds streets, N.E.).

KENILWORTH AQUATIC GARDENS

SOUTHWEST

The National Mall and West Potomac Park

LOCATION
U.S. Capitol to the Lincoln Memorial and south to the Tidal Basin
TELEPHONE
(202) 426-6841
WEBSITE
http://www.nps.gov/nama/index.htm
http://www.si.edu/Visit/Hours (to confirm museum hours)
SIZE
Many people say the area between the Lincoln Memorial and the Washington Monument is part of the National Mall. I certainly believed that for many years. In its literature, the National Park Service brings it all together as part of the "National Mall and Memorial Parks." But the original National Mall covers 300 acres from the U.S. Capitol west the Washington Monument. West Potomac Park, another 390 acres, continues the long, grassy rectangle, running from the Washington Monument west to the Lincoln Memorial. It also includes the Tidal Basin, which sits just south.
HABITATS
lawns, pools, tree groves, reservoir, and river's edge

Natural History   The area now occupied by the National Mall and West Potomac Park was once tidal mudflats, marsh, and forest including such species as sweetgum, sycamore, oak, hickory, black willow, and river birch. The Potomac River's banks sat along what is today the eastern edge of the Tidal Basin. Undoubtedly this habitat mosaic was rich in wildlife, including many species no longer found in the city.

Today, the area is completely transformed. The National Mall and West Potomac Park still attract wildlife, particularly birds and butterflies, drawn by the wide open green space and its location along the river. But virtually all the vegetation of this area was carefully planted to serve a certain function, be it to provide shade or frame a monument, or make some other landscape architecture or historical statement.

Human History   Each year, more than 25 million tourists visit the National Mall and West Potomac Park, making the area one of the world's most-visited green spaces. Most come to see the U.S. Capitol, 10 Smithsonian museums, the National Gallery of Art, the monuments, and the Yoshino cherry trees when they bloom. It is

Canada geese take a dip in the Reflecting Pool, between the
Washington Monument and the Lincoln Memorial.

hard to imagine a time when no
monuments stood, and no exotic cherry trees grew there.

But this area between the Fall Line and the confluence of the
Potomac and Anacostia Rivers provided rich fishing, hunting, and
farming for Algonquian-speaking Native Americans before and
during the arrival of Europeans.

The 1600s saw growing conflicts between Algonquian and
Iroquois tribes in the region, and between them and the growing
numbers of European settlers. Also, introduced diseases took their
toll on Native American communities. By the start of the 18th century, few remained in the area.

In the 1700s, settlers converted riverside areas to pasture and
cropland. Tobacco was an important crop. The area was rural but
settlements sprouted up, including the new port towns of Georgetown and Alexandria, both established mid century.

George Washington chose the capital site and French-born city
planner Pierre Charles L'Enfant was hired to lay out an impressive
city. In 1791, L'Enfant presented his plan. As part of his vision,
L'Enfant pictured a large open area stretching west from the U.S.
Capitol to the Potomac. Another component of L'Enfant's plan was a
city canal to provide boat access between the rivers. This Washington
City Canal, plagued by silting, gradually became an embarrassment,
part open sewer, part mosquito-breeding pit. It was filled in by the
1870s.

Construction of the Washington Monument began in 1848.
Today, at 555 feet tall, the monument's pointy top is now the city's

THE NATIONAL MALL

highest point. But it took a long time to reach that high. By 1858, funding petered out. For the next 18 years, the building stood at just 156 feet tall. During the Civil War, the uncompleted obelisk was the backdrop for a huge cattle yard. In 1876, President Ulysses S. Grant approved authorization for the federal government to resume construction. In 1885, almost 40 years after it was started, the Washington Monument was completed.

The area of West Potomac Park, where the Lincoln Memorial now sits, was once a flood-prone river bend lined with tidal marsh. In 1867, the U.S. Army Corps of Engineers was tasked to deal with the Potomac's periodic flooding, constant siltation problems, and poor drainage that impeded and embarrassed the capital city's residents. After studying the area, the corps received $50,000 in funding from Congress to begin clearing a channel and filling in the wetlands. Over the next four decades, dredge spoil from this ongoing project built up the shoreline to create dry land. In 1897, Congress authorized the conversion of much of this reclaimed area to parkland. This was, essentially, the birth of West and East Potomac Parks, although their official designations came later. There was still a lot of work to do. Although much of the land took shape well before, dredging, filling, and planting in these parks and on the west side of the National Mall were not fully complete until 1912.

Once dry land prevailed near the Potomac shore, a monument to bookend the Capitol on the other side of the Mall took shape. Although plans for a Lincoln memorial began shortly after the

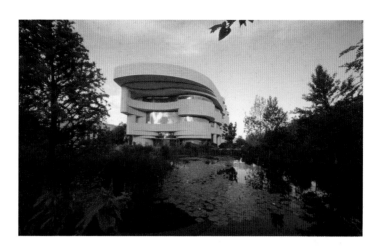

The National Museum of the American Indian is landscaped
with many native plants that attract wildlife.

president's assassination in 1865, the site was not chosen until 1901.
Construction began in 1914 and was completed in 1922. The Piccirilli
brothers of New York carved the 19-foot-high, 19-foot wide Lincoln
statue, which was assembled from 28 marble sections. The monument building is made of Colorado Yule marble and Indiana
limestone.

At a very brisk pace, you can stroll from the Lincoln Memorial,
past the Washington Monument, to the U.S. Capitol in a half hour.
This is a 2-mile walk through some of the city's most iconic and
stately parkland. (Note: During the summer, beat the heat by walking
as early as possible. Wildlife activity is at peak then, anyway.) West
Potomac Park and the National Mall form one of the country's most
famous rally and protest sites. On August 28, 1963, during Martin
Luther King, Jr.'s "I Have a Dream" speech, the view from the Lincoln Memorial steps east to the Washington Monument was wall-to-
wall people, except for the tall elms and the Reflecting Pool's interior.

The National Mall and West Potomac Park became part of the
National Park System in 1965. Monuments have since been added.
As part of bicentennial celebrations in 1976, Constitution Gardens
was created on land that had been occupied by military buildings
from World War I through 1970.

The city's quadrants are drawn out from the U.S. Capitol, which
straddles the Mall on its east end, facing the Washington Monument
to the west. The dividing line runs down the Mall's center, with the
north side in Northwest and the south side in Southwest. Constitution
Avenue runs along the Mall's northern boundary, while Independence

American elms tower over walkways lining the National Mall and West Potomac Park.

Avenue flanks its south side. West Potomac Park is adjacent to the Mall, and just below the cherry tree–flanked Tidal Basin is East Potomac Park. These parks can be reached by foot from the National Mall.

Main Walkways   For the naturalist, the National Mall is at its best well before the museums open. (Most Smithsonian museums open at 10 a.m., exceptions being the Smithsonian Castle, open by 8:30 a.m., and the American Art Museum and Portrait Gallery, open at 11:30 a.m.) Each day, sanitation workers collect and haul off three to four tons of garbage from the National Mall. Before it's taken away, this daily trash bounty attracts some wildlife, such as gray squirrels, ring-billed gulls, pigeons, starlings, grackles, and house sparrows, to benches and trash cans. These animals catch the eyes of birds of prey. Depending on luck and time of year, Mall raptors may include red-tailed, Cooper's, and sharp-shinned hawks. The Mall's open, waterside position is probably the greatest draw to raptors. You may see an osprey, bald eagle, or perhaps a falcon pass by. Eastern kingbirds nest in some of the tall shade trees by the U.S. Capitol and west, and in migration warblers, vireos, and other songbirds might show up during your early morning stroll.

According to the National Park Service, more than 9,000 trees grow on the National Mall. About 2,300 of these are American elms. These tall trees with candelabra-like branches line the National Mall and West Potomac Park leading to the Lincoln Memorial, on either side of the Reflecting Pool. Tree experts monitor them for Dutch elm disease, an introduced pathogen that decimated these trees in many areas.

The scene around the Smithsonian Castle is somewhat typical of the lavish landscaping around the museums and office buildings in the area: The front and back of the castle are flanked by large hollies, and southern magnolias and a large ginkgo are found in the back. In summer, purple coneflower, which attracts goldfinches, grows alongside many other flowers.

From mid-August to the first days of October, the lights around the U.S. Capitol and Washington Monument draw migrating common nighthawks. Dusk is the best time to see them. Also, on overcast days, a few of these birds may cruise overhead in erratic flight, looking like a cross between a swallow and a slow falcon. In May, nighthawks show up but in smaller numbers.

The Yoshino cherry trees growing around the Washington Monument were part of a gift from the Japanese government in 1965 when Lyndon B. Johnson was president and First Lady Lady Bird Johnson was involved in beautification projects around the city.

THE NATIONAL MALL

Enid A. Haupt and Mary Livingston Ripley Gardens The Enid A. Haupt Garden sits on more than four lush acres behind the Smithsonian Castle—and atop the subterranean National Museum of African Art, Arthur M. Sackler Gallery, and the S. Dillon Ripley Center. Landscaped with flowers, shrubs, and small trees and graced with many benches, this is a nice place to rest after a long walk between museums, as long as it's not a sweaty summer afternoon. The Mary Livingston Ripley Garden is next door, by the Arts and Industries Building. This park is similar but much smaller.

While sitting on a bench, you might see house finches (fond of the elaborate hanging baskets), northern mockingbirds, common grackles, American robins, and song sparrows. The shade-loving gray catbird nests here and in other nearby garden settings from spring to summer. Over the years, these small urban oases sheltered some out-of-place birds, including a Virginia rail, an ovenbird (in winter), and black-chinned and rufous hummingbirds.

National Museum of the American Indian This museum's grounds emulate pre-European landscapes—upland hardwood forest, croplands, meadow, and wetlands. More than 150 plant species grow here, including native oaks, pines, magnolias, pawpaw, cattail, and other plants rarely if ever planted on the National Mall.

Whether frosted with snow or tucked in its spring blanket of Yoshino cherry blooms, the Jefferson Memorial is a stunning sight. Ducks and gulls flock to the adjacent Tidal Basin.

Coneflowers and goldenrod draw goldfinches and butterflies. The man-made wetland east of the museum may host mallards, red-winged blackbirds, common yellowthroats, swamp sparrows, and some passing surprises, such as migrating palm warblers.

The Smithsonian Butterfly Habitat Garden This garden covers 11,000 square feet on the east side of the National Museum of Natural History, at 9th Street, N.W., just below Constitution Avenue, N.W. Native plants from wetland, meadow, and woodland edge grow here alongside more ornamental garden flowers. Signs explain which of these plants provide nectar to butterflies and which of them host eggs, larvae, and pupae. From spring through autumn, this space attracts a variety of butterflies, from mourning cloaks to red admirals, skippers, and monarchs. The nectar-providing plants also attract honeybees, including many from the colony maintained at the O. Orkin Insect Zoo, located on the National Museum of Natural History's second floor.

Next door, the National Museum of American History has two outside landscapes highlighting victory gardens and heirloom flower varieties.

Constitution Gardens   Constitution Gardens is located at the northwest corner of West Potomac Park, between the Lincoln Memorial to the west and the Washington Monument to the east. The man-made lake here adds a natural look to the park and is often a productive place to see wildlife. *Note:* This area will undergo renovations that may last a few years, which will affect wildlife that use the site, so please check National Park Service websites for updates. In the past, when the pond has been drained for maintenance, the exposed mud attracted migrating shorebirds.

Dedicated in 1976, Constitution Gardens honors the 56 men who signed the Declaration of Independence. Their names are on stones dotting the island in the lake's middle. To find birds here, it's best to arrive early, before the daily crowds. In spring and fall migration, watch for thrushes, thrashers, sparrows, and towhees, among others. From November to early May, the lake draws waterfowl. In addition to the usual mallards, watch for American coot, American wigeon, lesser scaup, ring-necked duck, wood duck, mergansers, and others (sometimes even redhead). Early in the day, migrating ducks may also mingle with local mallards and Canada geese in the long, rectangular Reflecting Pool, just below the gardens and between the Lincoln Memorial and Washington Monument.

Tidal Basin   This scenic, roughly clover-shaped body of water covers 107 acres and sits alongside the Potomac, ringed with flowering cherry trees and accented by the Jefferson and F.D.R. memorials. The basin, which averages 10 feet deep, rejuvenates itself by bringing in Potomac River water then flushing it into the Washington Channel. A paved 2-mile path circles the basin. At its south end, visitors cross the Inlet Bridge, where high tide water flows in from the Potomac to feed the basin. At low tide, water flows out to the Washington Channel via the Outlet Bridge, which visitors walk over east of the Jefferson Memorial. Between the memorial and the Outlet Bridge grows the indicator tree used to gauge the peak of Yoshino cherry tree blooming each year. Of unknown providence and type, this tree blooms about a week earlier than the Yoshinos.

The basin's northwest corner, just below Independence Avenue, S.W., has some of the oldest remaining Yoshino cherries and was where First Lady Helen H. Taft planted the first trees in 1912. Bronze plaques mark the spots. The 350-plus-year-old Japanese lantern that sits here has been lit each year since 1954 to start the National

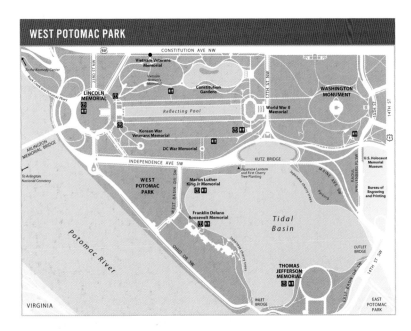

Cherry Blossom Festival. Just west of the Jefferson Memorial grows a stand of Usuzumi cherries, the oldest-known type of Japanese flowering cherry. They originate from a 1,400-year-old tree in Japan.

For naturalists, this tamed landscape still holds wildlife surprises. For example, from fall to spring, the Tidal Basin is one of the city's best gull- and waterfowl-watching spots. A spotting scope comes in handy here. Walk the periphery on the paved pathways, scanning the waters for loons, grebes, and a variety of ducks. Particularly in late afternoons during the winter, you may see hundreds to a thousand or more gulls. These birds roost here for the night, or in nearby East Potomac Park. A rare gull such as an Iceland or glaucous gull may turn up among them. In spring and fall, the graceful Bonaparte's gull and some Caspian terns fly past, and a few shorebirds may loaf along the water's edge or in rain pools.

Beneath the water swim largemouth bass, channel catfish, migrating striped bass, and other Potomac fishes.

Tidal Basin Cherry Trees From 1912 to 1920, many of the 1,800 Yoshino cherries given to the United States by Japan found a home at the Tidal Basin. The remainder, along with cherries of 11 other varieties, were planted next door in East Potomac Park (see separate entry). New trees are periodically planted to replace those that have died over the years. The National Arboretum and National Park Service work together to try to preserve the lineage of the few original trees, growing cuttings of surviving founder trees and then planting the resulting young trees. Today, there's a rebounding beaver population living along the Potomac, and periodically, these lumbering rodents appear at the Tidal Basin. In the past, some were relocated after damaging some beloved cherry trees.

The Yoshino blossoms usually peak very late in March or the first days of April.

To check cherry blossom bloom times and events, see the following websites: http://www.nps.gov/cherry/cherry-blossom-bloom.htm (National Park Service bloom schedule for cherry blossoms); http://www.nps.gov/cherry (Cherry Blossom Festival); http://www.nps.gov/cherry/cherry-blossom-maps.htm (maps); http://www.national cherry blossom festival.org/ (National Cherry Blossom Festival).

GETTING THERE

By Car   Parking along the periphery of the National Mall can be challenging, especially after 8 a.m. According to the National Park Service, each weekday, more than 440,000 vehicles pass through the Mall via Constitution and Independence Avenues. It takes patience to cruise these streets and others for free or metered spots. Arriving early is a good idea, for seeing the most wildlife, getting a parking spot, and in summer escaping the day's worst heat. Paid indoor parking garages are available within a few blocks of many parts of the National Mall.

Parking near the Tidal Basin is also limited. However, there are hundreds of spaces in adjacent East Potomac Park, although these often fill on warm weekend days. During cherry blossom season, a free "blossom shuttle" runs from East Potomac Park. It circles this park but also has a stop at the Jefferson Memorial welcome area, at the south edge of West Potomac Park.

Weekend parking along Constitution Avenue, N.W., near Constitution Gardens is often easier than along parts of the National Mall. During the week, both areas are busy.

By Metrorail and by Foot   Metrorail is an easy way to reach the National Mall. From late May to September, far more often than not, the Mall is a hot, humid place. Fortunately, there are plenty of benches, shady elms, and food, drink, and ice cream vendors. A daypack stocked with water, sunscreen, a hat or two, and a map will go a long way toward making your visit more comfortable. All of the National Mall and West Potomac Park can be explored on foot.

Smithsonian station, on the Blue and Orange Lines, is the only station with an exit onto the National Mall. The exit is near the

Smithsonian Castle and just across from the National Museum of Natural History and the National Museum of American History.

To reach Constitution Gardens, exit the Farragut West Metro station (Blue and Orange Lines) and walk eight blocks south on either 17th, 18th, or 19th Streets, N.W. All three dead end at Constitution Avenue, N.W. There are various paths into the park, which is right across Constitution Avenue, N.W. Another alternative is to exit Foggy Bottom station (also the Blue and Orange Lines) and head south on 23rd Street, N.W., to Constitution Avenue, N.W. Cross the street and enter the park. The Vietnam Veterans Memorial is here. Turn left once in the park to head toward Constitution Gardens.

To reach the Tidal Basin, walk just west of 17th Street, N.W., south of Constitution Gardens and the Reflecting Pool, and pass the World War II Memorial. You can also reach the Tidal Basin by walking west from the Smithsonian Metro station (Blue and Orange Lines), after leaving via the Independence Avenue exit.

To arrive between the museums on the eastern side of the National Mall, take the Blue and Orange Line Smithsonian station, the L'Enfant Plaza station (intersection of the Yellow and Green and Blue and Orange Lines), or the Blue and Orange Line station at Federal Center, S.W.

By Metrobus There are many options, depending on your starting point. See MetroOpensDoors.com to plan a trip and see routes.

By Bike There are many ways to cycle to the National Mall and West Potomac Park, including taking the Rock Creek and Potomac Parkway south through Rock Creek Park and then passing the John F. Kennedy Center for the Performing Arts before arriving at the Lincoln Memorial. Cyclists leaving Arlington, Virginia, can cross the Arlington Memorial Bridge from the Mount Vernon Trail to reach these parks.

East Potomac Park

LOCATION
This park is located on the east side of the Potomac River, just south of the Tidal Basin. The southern tip of the island, Hains Point, is where the Potomac and Anacostia Rivers and Washington Channel meet.

TELEPHONE
(202) 426-6841

WEBSITE
http://www.nps.gov/nama/index.htm

SIZE
327 acres

HABITATS
river, channel, golf course, lawns, tree groves

Natural History   The area that is now East Potomac Park was once tidal wetlands and the natural confluence of the Anacostia into the Potomac River. The area was likely rich in aquatic plant and bird life.

Human History   Mistaken by many for a peninsula, East Potomac Park is actually a dagger-like island made of fill dredged from the rivers over decades. In 1867, the U.S. Army Corps of Engineers took on the challenge of clearing a channel and filling in tidal wetlands and stagnant, sewage-plagued areas along the city's low-slung western shoreline. Periodic Potomac flooding, chronic siltation problems, and raw sewage were making the capital city an embarrassment. In 1882, the Corps began a concerted effort to contain soil, to channel water away from runoff areas, and to stem sewage flows. Until the early 1900s, fill generated by these efforts buried tidal wetlands and built up land that later became West Potomac Park, with its Tidal Basin, and then the peninsular East Potomac Park. In 1897, Congress declared much of this reclaimed area as parkland designated for city residents' recreation and relaxation, although the parks' formal designation came later.

Views, Cherry Trees, and Fishing   Like a curved dagger sheathed between the Potomac and Washington Channel, East Potomac Park tapers to a sharp tip at Hains Point. Looking south from the point, you see Ronald Reagan Washington National Airport and distant Alexandria, Virginia, on the right (west) bank of the Potomac and Joint Base Anacostia-Bolling on the left (east), with its north part facing the Anacostia River and its lower part the Potomac.

Along Ohio Drive, you find not only Yoshino but also the later-blooming Kwanzan and other flowering cherry varieties.

East Potomac Park is one of the city's most popular fishing spots and a favorite place to see the later-blooming Kwanzan cherry trees. Ornamental cherry and other trees line Ohio Drive, which skirts the island's perimeter, providing water views along most of its course. At the north end of the park, you can park and scan the adjacent Tidal Basin, just north.

A golf course, tennis courts, and a swimming pool occupy the park's northern midsection. Biking, walking, running, picnicking, photography, and birding are favorite pursuits. The Washington Channel runs along the park's east side, a sheltered, small water body flushed by the Tidal Basin. This is a favorite fishing spot. Anglers come here to catch, among other species, common carp and largemouth bass.

East Potomac Park is where park personnel planted the *other* cherry trees sent by the government of Japan in 1912. Two weeks after the Tidal Basin's Yoshino cherry blooms peak, the Kwanzan cherries along the west side of East Potomac Park come into their glory. The Kwanzan flowers are pink and bloom in dense clusters. Some of the famed white Yoshino cherries are here as well, lining both sides of the island along Ohio Drive, and small numbers of other types include Takesimensis, Akebono, weeping Japanese cherry, and Sargent or mountain cherry.

Birding Tips　Migration and winter are "high season" for those interested in East Potomac Park's birdlife. From late fall to early

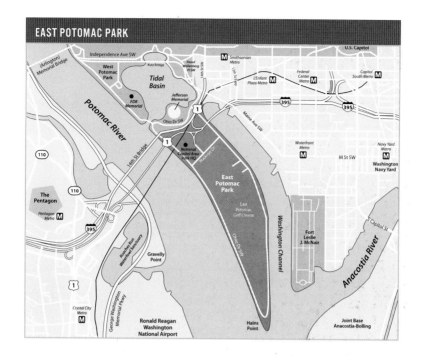

Map labels:
(Arlington) Memorial Bridge · Independence Ave SW · West Potomac Park · Kutz Bridge · Raoul Wallenberg Pl SW · 14th St SW · Smithsonian Metro · U.S. Capitol · 12th St SW · Federal Center Metro · Capitol South Metro · L'Enfant Plaza Metro · Tidal Basin · FDR Memorial · Jefferson Memorial · Ohio Dr SW · Potomac River · 110 · The Pentagon · Pentagon Metro · 395 · 1 · 1 · 110 · National Capital Area Parks HQ · 14th St Bridge · Maine Ave SW · 395 · 395 · Waterfront Metro · M St SW · Navy Yard Metro · Washington Navy Yard · East Potomac Park · East Potomac Golf Course · Washington Channel · Fort Leslie J. McNair · Capitol St · Anacostia River · Roaches Run Waterfowl Sanctuary · Gravelly Point · George Washington Memorial Pkwy · 1 · Crystal City Metro · Ronald Reagan Washington National Airport · Hains Point · Joint Base Anacostia-Bolling

spring, rafts of lesser scaup and other bay ducks abound, common loons and grebes regularly appear, and the city's largest gull concentrations—sometimes in the thousands—occur here. Local birders keep an eye on winter weather. When extended cold snaps partially freeze the channel and nearby rivers, fish kills may occur. These events draw the largest numbers of gulls. Among the many ring-billed, herring, and great black-backed gulls, diligent observers often spot a few lesser black-backed gulls and sometimes rare species such as Iceland and glaucous gulls. Late-afternoon numbers are higher because the birds congregate in the area to spend the night. In late summer, you may see Forster's and Caspian terns (also seen in spring) and laughing gulls, before they head south for the winter. Bonaparte's gulls show up in April.

A few raptors usually preside over the avian congregation. Bald eagles pass over the river, sometimes perching on tree snags. Ospreys nest on the railroad bridge paralleling the 14th Street Bridge where it crosses the Potomac, viewed from the park's west side. Many birders consider Hains Point the city's best falcon-watching spot. Small numbers of peregrine falcons, merlins, and kestrels appear during migration and sometimes in winter.

In the park's northeast corner, you can often see a few roosting black-crowned night-herons. Best viewing is from the shoreline

Many birders visit East Potomac Park, where they might be overheard saying, "Look, there's a first-winter ring-billed gull!"

looking north over the top of the Washington Channel, where you may spot the birds perched in trees flanking the 14th Street railroad bridge.

During migration, swallows pass through, and northern rough-winged and barn swallows nest nearby. After rains, migrating shorebirds and American pipits sometimes show up at puddles and in muddy areas in the golf course and elsewhere. Spring to fall, Baltimore and orchard orioles, warbling vireos, and yellow warblers sing or hunt insects in the trees lining Ohio Drive.

To enjoy the birds and serenity of the spot, plan to arrive early. Parking lots may fill and close in afternoons, when peak crowds are there. During "bad" weather days, the park is less crowded and more likely to have birds because storms discourage golfers and picnickers but drive soggy, wind-blown migrants to terra firma.

GETTING THERE

By Car   If you arrive from Northern Virginia on I-395, take the East Potomac Park exit just after crossing the 14th Street Bridge into the city.

From the west side of the city, follow Rock Creek and then the Potomac's edge along the Rock Creek and Potomac Parkway, headed southeast. Continue straight onto Ohio Drive, S.W., then across the Tidal Basin's Inlet Bridge to the park.

From just east of the White House, take 15th Street, S.W., south until it changes into Raoul Wallenberg Place, S.W. Continue south, following the edge of the Tidal Basin clock-wise then across the

Outlet Bridge to a sign-posted left turn onto Ohio Drive, S.W., and East Potomac Park.

Although Ohio Drive, S.W., is lined with hundreds of free parking spaces, they may fill fairly early on warm-weather weekend days. Parking lots may close on crowded afternoons. Also, check local listings to make sure the park road isn't closed for races or other events.

By Metrorail The nearest Metrorail station is Smithsonian, on the Blue and Orange Lines. From the Independence Avenue exit, walk west on Independence Avenue, S.W. Cross 14th street, S.W., then at the next block turn left (south) on Raoul

East Potomac Park draws many raptors, including this bald eagle.

Wallenberg Place, S.W. On the left, you pass the U.S. Holocaust Memorial Museum and the Bureau of Engraving and Printing before you reach the Tidal Basin. Walk a bit southeast around the Tidal Basin, passing over the small Outlet Bridge, then turn left (south) at the sign to enter the park.

By Metrobus See MetroOpensDoors.com to check current bus routes.

By Foot or by Bike See "By Metrorail" directions above. From West Potomac Park, you reach East Potomac Park by walking around the Tidal Basin. Along the basin's west side you cross following Ohio Drive via the Inlet Bridge. From the basin's east side you enter via sidewalks that pass beneath the 14th Street overpass to Ohio Drive's east side.

Lady Bird Johnson Park and Lyndon Baines Johnson Memorial Grove

LOCATION
on the George Washington Memorial Parkway, between I-395 and the Arlington Memorial Bridge

TELEPHONE
(703) 289-2500

WEBSITES
http://www.nps.gov/lyba/ and www.nps.gov/gwmp

SIZE
121 acres, including the 17-acre memorial grove

HABITATS
river shore, gardens, groves with planted shrubs and trees

Natural History   Lady Bird Johnson Park sits on soil dredged out of the Potomac River in 1916. The first lady picked the site for its views of the Washington skyline across the river. This man-made habitat serves not only the public but also some wildlife living along the Potomac River. The park is lined with oaks, maples, weeping willows, and dogwoods. In early spring, more than 10,000 tulips and a million daffodils blanket parts of the park.

Human History   Although it looks like part of the Virginia shoreline, the park sits within Washington, D.C. Located on Columbia Island, it was dedicated as Lady Bird Johnson Park in 1968, in honor of a first lady dedicated to beautifying the nation's capital and the country's roadways and public spaces.

The Navy and Marine Memorial, a gull flock sculpture, is one of the most prominent landmarks seen by motorists driving past the park on the George Washington Memorial Parkway.

The Shoreline   With a backdrop of Washington's scenic monumental skyline, this tranquil, flower-filled park sits on a D.C. island sandwiched between Arlington National Cemetery and the Virginia shoreline to the west and busy car and bicycle corridors flanking the Potomac River on its east side. The river attracts waterfowl, gulls, great blue herons, ospreys, and double-crested cormorants. The narrow Boundary Channel, on the island's west side, offers quieter waters where a visitor might find roosting black-crowned night-herons, a belted kingfisher, or other surprises.

Lyndon Baines Johnson Memorial Grove   Located on 17 acres within the park, this memorial to the 36th president is a good place

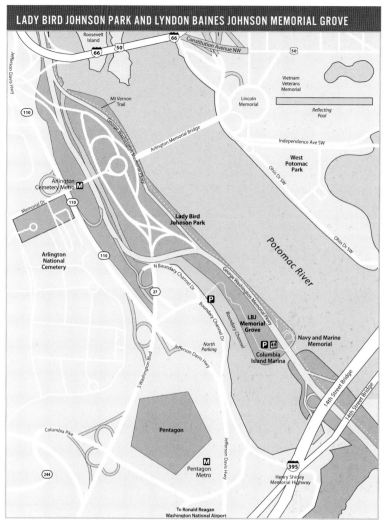

for visitors to enjoy nature. A 19-foot-tall granite monolith dedicated to Johnson is surrounded by white pines. The stone was quarried in Texas 35 miles from the LBJ ranch. Other plants at the grove include flowering dogwood, azaleas, rhododendron, and various perennials.

The grove was dedicated in 1976, three years after Johnson's death. Now about 40 years old, the trees are tall and attract migrating and wintering songbirds. Pine warbler, winter wren, brown creeper, hermit thrush, and golden-crowned kinglet often occur here, sometimes joined by orange-crowned warbler, pine siskin, or red-breasted nuthatch, birds scarcely seen in the city.

A granite monolith dedicated to President Lyndon Baines Johnson
sits among the tranquil white pines.

Leaf litter and pinecones at the memorial grove.

GETTING THERE

By Car The park is located on the George Washington
Memorial Parkway between I-66 (Theodore Roosevelt Memorial
Bridge) and I-395 (14th Street or George Mason Bridge). It can also
be reached by the Memorial (Arlington Memorial) Bridge, which
runs west from the Lincoln Memorial, across the Potomac River to
the park. Boundary Channel Drive, near the Pentagon, also provides
access. From East Potomac Park in D.C., you can take the George

Mason Memorial Bridge (I-395) across the river. Cross the George Washington Memorial Parkway and take the Boundary Drive exit. Off this exit, turn right, then follow signs.

By Metrorail The closest Metrorail station is the Blue Line stop at Arlington Cemetery, from which visitors can walk east on Memorial Drive into the park, then take the Mount Vernon Trail south to the memorial grove.

By Foot or by Bike The 18.5-mile Mount Vernon Trail passes through the park on its way along the Potomac shoreline.

NEARBY

The National Park Service administers the 25-mile-long George Washington Memorial Parkway, which runs through the park. Some of the area's best birding spots are farther down this road. These Virginia sites include Gravelly Point, at the north edge of Ronald Reagan Washington National Airport, where you look out onto Washington, D.C., waters. Directly across the parkway from Gravelly Point is the Roaches Run Waterfowl Sanctuary. Here you may see green herons and ospreys in spring and summer and small numbers of waterfowl, including pied-billed grebes and hooded mergansers, in fall and winter.

Continuing south on the parkway, you reach Daingerfield Island (Washington Sailing Marina), then the following sites south of the Virginia city of Alexandria: Jones Point Lighthouse, Belle Haven Park and Dyke Marsh Preserve, Fort Hunt Park, Riverside Park, and finally Mount Vernon, George Washington's home and burial site. Ospreys, bald eagles, ducks, double-crested cormorants, and great blue herons are among the most noticeable birds along this route. Baltimore and orchard orioles and warbling vireos nest in the parkway trees.

Dyke Marsh Preserve is of particular interest to the naturalist. This is one of the area's largest freshwater tidal marshes, where you may find beavers, muskrats, cattails, wild rice, and an assortment of amphibians and reptiles. More than 250 bird species have been recorded. At low tide, the flats off adjacent Belle Haven Park, especially at its north tip at Hunting Creek, provide one of the best shorebird viewing areas near the city.

SOUTHEAST

Anacostia Park

LOCATION

About 5 miles along both sides of the Anacostia River, from the Washington, D.C. / Maryland line southwest to the Frederick Douglass Memorial Bridge.

TELEPHONE

(202) 472-3884

WEBSITE

Anacostia Park:

http://www.nps.gov/anac/

Anacostia Riverwalk Trail:

http://www.anacostiawaterfront.org/awi-documents/anacostia-riverwalk-trail-documents/dc-anacostia-riverwalk-trail-map/

SIZE

More than 1,200 acres, including Kenilworth Aquatic Gardens (see separate account).

HABITATS

river, shoreline, and some mudflats, lawn, shrubby areas, meadow, floodplain forest

Natural History The area that is now the park was once a quarter-mile-wide wetland of wild rice and other native vegetation. In the 1920s and 1930s, dredge spoil filled the wetland areas to build up dry land and parks. Today, Anacostia Park is a long ribbon of parkland that straddles both sides of the Anacostia River. Although much of the park's lower part consists of athletic fields and other recreation facilities, the property here is dotted with some wild patches of meadow and scrub. Along the river's edge, the National Park Service is re-establishing native vegetation. Above the park's midsection, north of the John Phillip Sousa (Pennsylvania Avenue) Bridge and nearby railroad bridges, floodplain forest cloaks the riverbanks northeast to the Maryland line.

Along the river, watch the trees in spring and early summer for nesting Baltimore and orchard orioles, eastern kingbirds, and warbling vireos. Barn and northern rough-winged swallows wheel over the river and grassy recreation fields, snapping up insects. Although seldom seen, red and gray foxes prowl parts of the park. Monarch butterflies often show up, particularly during their fall migration.

Exposed tidal mudflats at Anacostia Park attract waterbirds.

Gulls, cormorants, ducks, and geese may be seen from either side of the John Phillip Sousa Bridge and from the shore overlooking the park's southern limit, where the Frederick Douglass Memorial Bridge (South Capitol Street) crosses to the Anacostia River's east bank. In spring and summer, ospreys nest on a pier beneath this bridge and most of the year, gulls rest on the pilings. Inside the park along the river shore, watch for ring-billed, herring, and great black-backed gulls. A lesser black-backed gull or two sometimes shows up in these gull aggregations. From August through fall, laughing gulls and Forster's terns loaf along the shoreline, and Bonaparte's gulls and Caspian terns put in appearances during migration. Canada goose flocks should be scanned for rarities such as cackling and white-fronted geese. Killdeer and other shorebirds often turn up on exposed mudflats in grassy areas or rain pools. Over the years, an impressive number of rare birds have been recorded in this park, particularly during migration.

During migration and winter, waterfowl diversity can be impressive and changeable, depending on weather and water levels. Watch for ruddy duck, bufflehead, horned grebes, and others. Any unkempt, shrubby areas should be checked for common yellowthroat, yellow warbler, willow flycatcher, and sparrows. Weedy ditches may shelter wintering Wilson's snipe.

Raptors can turn up anywhere in this riverside park. They include American kestrel, peregrine falcon, red-shouldered and red-tailed hawks, and bald eagles, which can be seen along the river. March to October, osprey are present, and migration periods and

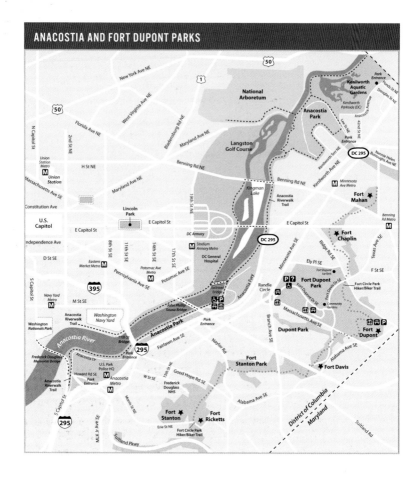

winter bring merlin, which hunt a wide variety of smaller birds. Raptors sometimes use the playing field goalposts for perches.

Human History The park and area of the city get their name from the Nacotchtank, or Anacostan, Algonquian-speaking Native Americans who inhabited, hunted, and fished the area. Their largest village, Nacotchtank, was located on the east bank of the Anacostia River close to where the park sits today. In 1608, English explorer Captain John Smith cruised up the tidal Potomac and met some Nacotchtanks, who he reported to be peaceful. As exploration and settlement moved north from James-town, English settlers developed land along the river. The new col-onists later both traded with and fought with the Nacotchtanks. The Nacotchtank community shrank because of disease, conflicts with colonists and other tribes,

Efforts to restore the Anacostia River include planting native trees along the shoreline in Anacostia Park and elsewhere.

and dispersal to other tribal areas. By the mid-1600s, perhaps a quarter of the population remained. The Nacotchtank community had disbanded by the close of the 1600s.

After the Revolutionary War, George Washington chose the meeting place of the Potomac and Anacostia Rivers as the site for the new country's capital city. In 1790, Congress approved the site that would be the capital of the new nation and, Washington hoped, a major port as well. At the time, tobacco was the most important crop grown in the area, planted and harvested on huge slave-worked plantations. By the late 1700s, after chronic problems with silt build-up, plans to make the Anacostia an important commercial waterway lost steam.

In 1799, the Washington Navy Yard opened, across from what is now Anacostia Park's west end and Poplar Point. This new facility sparked a local boom in residential development. At first a ship-building center, the Washington Navy Yard is still a Navy property, but today it is home to the Chief of Naval Operations, many naval commands, and is the headquarters for the Naval Historical Center.

Wetlands once flanked the Anacostia River, but almost all were buried during the 1920s and 1930s. The goal at the time—to develop dry, landscaped parks like West and East Potomac parks. In 1938, Congress purchased acreage for the park.

Today, Anacostia Park is one of the city's largest green spaces, following both banks of the river with which it shares its name. In its lower half, Anacostia Drive parallels the river along its east bank,

Open parkland with large scattered trees; a fine place for nesting orioles.

from the park's south end at the Frederick Douglass Memorial Bridge (South Capitol Street) to its middle, at a large recreational complex with playground, a roller skating pavilion, restrooms, an aquatic education center, and a boat ramp overlooking the river. Across the river on the river's west bank above Benning Road is the 18-hole Langston Golf Course.

The Anacostia Riverwalk Trail is a continuous 20-mile hiker/biker trail linking parkland and communities on both sides of the Anacostia River thanks to several river crossings, including South Capitol Street, East Capitol Street, and Benning Road. Another crossing is at railroad bridges just north of the John Phillip Sousa (Pennsylvania Avenue) Bridge and below East Capitol Street. Ospreys sometimes nest on one of the railroad bridges. The Anacostia Riverwalk Trail makes many parts of Anacostia Park more accessible for nature study. (See "By Bike" below.)

GETTING THERE

By Car   There are four main entrances to this long park and one to Kenilworth Aquatic Gardens (described in a separate section).

One entrance is just east of the Frederick Douglass Memorial Bridge (South Capitol Street) off Howard Road, S.E.

Another entrance is off Good Hope Road, S.E., just west of the Anacostia Freeway (295) and the 11th Street bridges.

To reach the northeastern Kenilworth Recreation Park section of the park, the entrance is just on the west side of the Nanny Helen Burroughs Avenue / Kenilworth Avenue (DC 295) interchange. The entrance to Kenilworth Aquatic Gardens (see separate entry) is just north of this entrance.

Another entrance is off Nicholson Street, S.E., next to the Anacostia Freeway (DC 295) and Pennsylvania Avenue (John Phillip Sousa) Bridge interchange. Here are driving directions from the city, coming from the north, entering the park at the Nicholson Street, S.E., entrance:

From south of the Capitol and the National Mall, take I-395, the Southwest / Southeast Freeway, to the John Phillip Sousa (Pennsylvania Avenue) Bridge. As you cross the bridge, stay in the right lane, then make the first right on Fairlawn Avenue, S.E., followed by another immediate right on Nicholson Street, S.E. (There is a park sign here.) Go under the DC-295 (Kenilworth Avenue) interchange and you will soon reach a T junction in the park at Anacostia Avenue and the river. You can go either way, south or north from there. (Note: The park road is one way on summer weekends.)

Visitors should plan to bag and take their trash home with them.

By Metrorail   If taking Metrorail, you reach the southwestern part of the park via the Green Line Anacostia station. Exit toward the Metro parking lot. Then take the sidewalk path and enter the park via a gate in the fence.

From the Potomac Avenue station on the Blue and Orange Lines, you can reach the midsection of the park by walking south across the Anacostia River on the Pennsylvania Avenue bridge on a sidewalk on the right (south) side. A paved path with steps leads into the park.

A number of buses pass by the parks' entrances. For current bus route information, see MetroOpensDoors.com.

By Bike   The 20-mile-long Anacostia Riverwalk Trail currently runs through the lower part of Anacostia Park on both sides of the river. When completed in 2015, it will pass through the park's upper section as well, including Kenilworth Aquatic Gardens, and will link to trails in Maryland.

For a map and updates on this growing trail, see the Anacostia Riverwalk Trail website listed at the beginning of this Anacostia Park account.

NEARBY

Frederick Douglass National Historic Site   Frederick Douglass National Historic Site (Cedar Hill) is located at 1411 W Street, S.E., at the corner of 15th and W Streets, S.E. This property sits about a half mile east of Anacostia Park's two southwestern entrances, on a hill overlooking the city. The famed abolitionist and friend of Abraham Lincoln lived here from 1877 and died here in 1895. In his diary, he wrote of a large eastern white oak tree in his yard. The tree grew northeast of his porch, and he likely saw it every day he lived there. Now 100 feet tall, this "witness" tree can be visited along with the house at this National Park Service property.

Fort Stanton Park   This large, forested park sits east of the Frederick Douglass National Historic Site. The earthworks for the Civil War–era fort built here lie in the park's southwest corner, behind Our Lady of Perpetual Help Catholic Church. Fort Stanton was built to protect the Washington Navy Yard. A large part of the park is forested, with eastern white oak, tuliptree, and black locust among the common tree species.

Fort Dupont Park

LOCATION
3600 F Street, S.E., Washington, DC 20019
(This park is shown on a map in the previous account, Anacostia Park.)

TELEPHONE
(202) 426-7723

WEBSITE
http://www.nps.gov/fodu/planyourvisit/directions.htm

SIZE
361 acres

HABITATS
forest, shrubby areas, community gardens, outdoor theater, and playing fields

Natural History Fort Dupont Park is one of the largest protected forests in the city. Among the rich flora here are widespread mountain laurel shrubs, which put on quite a show when they flower beneath the tree canopy from May to June. Also look for wild azalea, blueberry, various ferns, spicebush, and such locally rare plants as Indian pipe and cranefly orchid. Animals include such "out in the country" species as eastern box turtle and gray fox. Visitors regularly spot black rat snakes and white-tailed deer, as well as a profusion of birds.

Many trees here are 65 years or older, and the oaks produce prodigious amounts of acorns. This may help explain why Fort Dupont Park is the city's best spot for viewing wild turkeys. Also, the park's high points offer good raptor-viewing opportunities. Red-shouldered hawks, black and turkey vultures, migrating kestrels, and others are seen there.

This large forest park provides important nesting habitat for D.C.'s official bird, the wood thrush, along with ovenbird, scarlet tanager, red-eyed vireo, and other songbirds. Spring to summer, brushy areas, including one just inside the park at the Ridge Road, S.E. entrance, shelter blue grosbeaks and indigo buntings.

Human History Although very little evidence of it remains, Fort Dupont was once part of a chain of 68 forts protecting Washington from Confederate invasion during the Civil War. The fort stood in a hilly area with good sight lines over the otherwise low Coastal Plain terrain. Remnants of the earthwork can still be seen near the picnic area in the southeast part of the park, off Alabama

Fort Dupont Park protects a large block of mature urban forest.

Avenue. A moat and felled trees protected the six-sided structure. Like other forts circling the city, Fort Dupont had a sweeping view of adjacent farmland. Today, the area is urban. The fort saw no action during the war but sheltered freedom-seeking slaves working their way to Washington. The structure was named for Rear Admiral Samuel Francis Dupont, a naval commander famous for the November 1861 Union victory at Port Royal, South Carolina.

The National Capital Planning Commission acquired the area in 1930. In the northern part of the park, a golf course was built. This facility closed in 1970. Soon after, an indoor ice rink and playing fields were built, while the rest of the abandoned golf course was reclaimed by forest. This younger woodland contrasts the more mature acreage found further east and south in the park. At the park's core sits a large patchwork of community gardens. This is a nice area to watch butterflies and open-area birds.

The park has five principal trails that pass through all these habitats. All are marked. Plan your route ahead of time, as few loop options are possible. The trails can be viewed at

http://www.nps.gov/fodu/planyourvisit/upload/Map-with-GPS-trail-system.jpg.

GETTING THERE

By Bike The Fort Circle Park Hiker-Biker Trail starts 3 miles to the north, just south of Fort Mahan Park and just below Benning

The soldiers and horses are gone, replaced by the ovenbird's song, the blaze of blooming mountain laurel, and tranquil picnic grounds.

Road, just west of the Benning Road Metrorail station (Blue Line). It crosses East Capital Street, S.E., goes through Fort Chaplin Park, then leads through the east side of Fort Dupont Park and continues through parkland south to Fort Davis and Fort Stanton, before ending at the site of Fort Ricketts. A brochure showing the hiker-biker trail can be seen at http://www.nps.gov/fodu/planyourvisit/brochures.htm.

By Metrorail The Benning Road station on the Blue Line is the closest Metrorail station. From the station, it is a 1.5-mile walk down Texas Avenue, S.E., to the park entrance, where Texas Avenue, S.E., becomes Fort Davis Drive, S.E. Or you can walk almost half a mile to Fort Chaplin Park on East Capitol Street, S.E., and hike another six-tenths of a mile down the Fort Circle Park Trail to the park boundary of Fort Dupont Park. From there, you have several trail options.

By Metrobus During the week, the V7 and U2 buses reach the area near the park. On weekends, the V8 bus serves this area. The V7 and V8 buses can be caught at the Minnestoa Avenue Metrorail station on the Orange Line. Check the WMATA website at www.wmata.com or MetroOpensDoors.com to confirm current bus routes and to plan your trip.

By Car The park's vehicle entrances are at Fort Davis Drive, S.E., and Ridge Road, S.E.; Fort Davis Drive and Massachusetts

Avenue, S.E.; and Randle Circle, S.E., and Fort Dupont Drive, S.E. (where the park's activity center is located).

From downtown or from Virginia and I-395 North, take Pennsylvania Avenue, S.E., east, crossing the Anacostia River and Anacostia Freeway (DC 295) before reaching Minnesota Avenue, S.E. Turn left on Minnesota Avenue, S.E. Drive a half-mile and turn right onto Randle Circle, S.E., watching for the park sign. Enter the park at Fort Dupont Drive, S.E., the third "spoke" off the circle. Once in the park on Fort Dupont Drive, S.E., watch for the park's activity center, parking area, and restrooms, located just before the intersection with Fort Davis Drive.

From Prince George's County in Maryland and the east side of Washington, D.C., take Pennsylvania Avenue west to Minnesota Avenue, S.E., and turn right. Travel a half-mile and turn right onto Randle Circle, S.E., entering the park at Fort Dupont Drive, S.E., the third "spoke" off the circle. Once in the park on Fort Dupont Drive, watch for the park's activity center, parking area, and restrooms, located before the intersection with Fort Davis Drive.

Normally, the only available parking area is the lot at the activity center. You may be ticketed parking in other areas. The only exception is during summer outdoor concerts, when an F Street, S.E., lawn parking area is available as well. Call the park if you have questions, or you can see a parking map at http://www.nps.gov/fodu/planyourvisit/maps.htm.

4

animals

INVERTEBRATES   *Annelid*

Earthworm: *Lumbricus terrestris*

ETYMOLOGY
Lumbricus: worm; *terrestris*: land

Description   Pink, slimy to the touch, and ringed with many rib-like segments, the earthworm is a familiar sight to anyone who has worked the soil or walked a sidewalk during or after a rain. The only part of the earthworm that is not ringed is the wide band near the tapered front end of the worm, called the clitellum.

Lumbricus terrestris is the "night crawler" familiar as fishing bait. It originated from Europe and is commonly encountered in soils throughout the city. It grows up to a foot long, and its body is divided into about 180 segments. Each of the earthworm's segments holds four pairs of stiff, short bristles, which enable the animal to move through the soil and to hold fast in its burrow if attacked.

Common Locations   Sometimes flushed out of hiding by sudden rains or excessive moisture in the soil, earthworms face peril on sidewalks, where they may easily dry up and die. They spend most of their lives belowground, working through the soil.

Notes of Interest   Introduced European earthworms, which now dominate the soil in many areas, including in Washington, D.C., probably altered the nutrient dynamics of the soil in local parks and gardens.

Ecological Role   Earthworms feed on decaying organic matter in soil. Through their constant feeding and burrowing, they aerate the soil, allowing moisture in and facilitating root growth. Earthworms swallow soil while burrowing and as a way of feeding. Organic matter is digested, but the bulk of the soil passes through them. The resulting castings enrich the soil. Earthworms provide important food for birds such as American robins, moles, foxes, snakes, turtles, and frogs.

KEY POINTS
- Earthworm castings can be seen above ground at the entrance of worm burrows as little piles of tiny, mud-like balls.
- The slimy cuticle surrounding the earthworm protects it and allows the animal to breathe. An earthworm quickly dies if its cuticle dries out.
- Earthworms are hermaphrodites, carrying both eggs and sperm. When two earthworms mate, they exchange reproductive materials. Worms do not fertilize themselves.

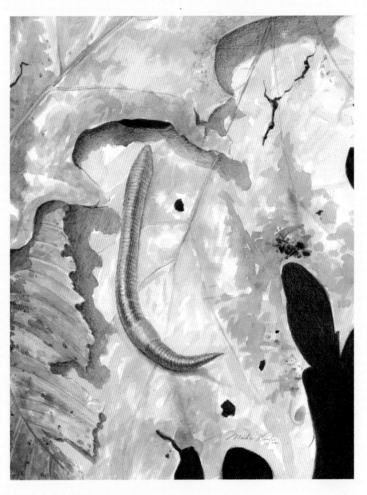

Plate 1

EARTHWORM

Brown Daddy Longlegs, or Brown Harvestman: *Phalangium opilio*

ETYMOLOGY

Phalangium: Greek for spider; *opilio*: Latin for shepherd

Description   The rust-colored, oblong body of the brown daddy longlegs, or brown harvestman, seems to float over the ground, transported by eight incredibly long, spindly legs. Most people mistake this leggy creature for a spider, but harvestmen belong to a separate arachnid order, Opiliones. Some differences: Harvestmen do not spin webs, but they do get caught in spider webs. They have at most two eyes, not eight like spiders. They also do not produce venom. They have what appears to be a single body segment (even though it is actually two segments fused together). Spiders normally have a distinct cephalothorax and abdomen.

Common Locations   Brown daddy longlegs are abundant at the base of tree trunks, on the forest floor, under logs, in shady spots in gardens and around the outside of houses. They are native to Europe but are now found throughout North America.

Notes of Interest   Cellar spiders or daddy longlegs spiders (*Pholcus phalangioides*) are true spiders that live in cellars and garages. Also named for their spindly legs, they spin webs and have two easily seen body segments.

Ecological Role   Daddy longlegs eat both decaying plant and animal matter as well as small live creatures such as aphids and other agricultural pests. Birds, spiders, and other predators eat them.

KEY POINTS

- The daddy longlegs does not bite and children often play with them.

Plate 2

BROWN DADDY LONGLEGS, OR BROWN HARVESTMAN

Goldenrod Crab Spider: *Misumena vatia*

Description   Bright yellow or white coloration hides this small predator among the blossoms. Female goldenrod crab spiders can slowly change color between yellow and white. This gender is most often seen because females are larger—almost a half-inch long—and more colorful. Males are tiny and rust red with whitish abdomens. With legs held to the side crab-like, goldenrod crab spiders grab unsuspecting prey often larger than themselves, injecting them with venom before feeding on them. They do not spin webs for catching prey; instead, they hide amid the flowers and pounce. These spiders sometimes make unplanned house visits when people bring garden-cut flowers inside for the dinner table. Their bite is not harmful to humans.

Ecological Role   Hidden among the flowers, goldenrod crab spiders capture a variety of garden insects, including flies and bees. When found, they are eaten by birds. They are active spring to early fall.

Rabid Wolf Spider: *Rabidosa rabida*

Description   Resembling a small tarantula, the rabid wolf spider is unjustifiably feared by people. Once thought to hunt in groups (thus, the wolf moniker) these spiders are solitary. The bite contains weak venom that may cause swelling but is not dangerous to people. The rabid wolf spider does not spin hanging webs. Instead, it burrows into the ground and spins silk to line its burrow. For northeastern spiders, wolf spiders grow large—the larger female's body may be an inch or more long. If measured with the legs, these animals can measure 4 inches across. Males are far smaller, with a body size of about a half inch.

Ecological Role   Rabid wolf spiders hunt a wide variety of insects, including crickets and cockroaches. In turn, they are eaten by wasps, birds, five-lined skinks, frogs, and other animals.

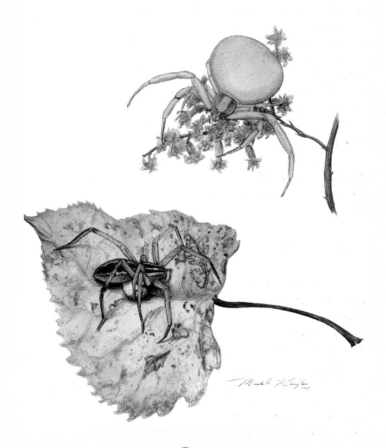

Plate 3

From top: GOLDENROD CRAB SPIDER & RABID WOLF SPIDER

Garden Millipede: *Oxidus gracilis*

Description   This small (up to three-fourth inch), many seg-
mented denizen of rich humus is often found under rocks or other
debris, curled into a tight coil. Despite the name, millipedes do not
have a thousand legs. Around the world, there are many species,
and the maximum leg count on one species reaches more than 750.
Most millipedes, however, have about 100 legs. The hard, shiny
cuticle protects the millipede as it burrows. Female millipedes lay
eggs in a little mound of dirt, guarding them until they hatch.

Ecological Role   Millipedes move slowly, using their many
legs to burrow into the soil. They eat soil and decomposing plant
matter and thus help in breaking down particles to form soil. Kids
love to pick them up. They do not bite but do exude a tennis-ball-
like odor as a defense mechanism.

Garden Centipede: *Lithobius forficatus*

Description   Flat, shiny, and orange, the fast-moving garden
centipede has 15 body segments and 15 pairs of legs, the front pair of
which is enlarged and used to capture and inject venom into prey.
Look for this up to 2-inch-long creature under bark, rocks, rotting
logs, and in the leaf litter of gardens and parks. Centipedes have
more distinct heads and far fewer legs than millipedes. They also
move much faster. They usually have back legs that resemble
antennae, giving the centipede an almost two-headed aspect. If
grabbed, this creature bites, although human skin is too tough to
penetrate. In any case, the sting is not dangerous.

Ecological Role   The garden centipede attacks any small
invertebrates it can catch, including earthworms, snails, and soft-
bodied insects. Now found throughout North America, the garden
centipede was originally a European species.

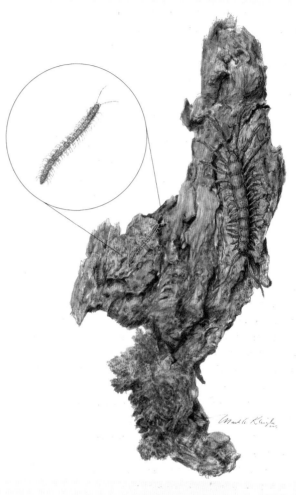

Plate 4
From left: GARDEN MILLIPEDE & GARDEN CENTIPEDE

Common Eastern Firefly: *Photinus pyralis*

Description   This beloved flashing insect is not a fly at all, but a soft-bodied beetle. The common eastern firefly, or big dipper firefly, has a flattened, oblong body with dull black wings and a red head dotted with an inky black spot. In summer, this insect is a familiar sight, flashing over lawns, parks, and gardens at dusk and after nightfall. The male performs a J-shaped flight, his abdomen flashing an intoxicating yellow green. Light-producing organs low on the abdomen produce the flash.

Ecological Role   Firefly larvae live underground or under debris, feeding on other invertebrates, including earthworms.

Lady Beetle: *Coccinella, Adalia,* and *Harmonia* spp.

Description   These beloved, glossy-shelled beetles are also called ladybugs, though they are not true bugs from an entomological perspective. The playing field has changed for ladybugs in recent years, with native species being replaced by introduced exotic species in many areas. A disappearing native species (*Coccinella novemnotata*) has nine black spots and another, two-spotted native (*Adalia bipunctata*) is declining in some areas and being introduced in others as biological control predators for farms and gardens. Meanwhile, the multicolored Asian lady beetle (*Harmonia axyridis*) has become abundant in many areas after its introduction from Japan, also as a biological control. Large numbers enter some homes to escape the cold. Markings of individuals in this species vary greatly in color and spotting. Another increasing introduced species (*C. septempunctata*) has seven spots and is red. Ladybugs keep their thin-membraned flight wings folded beneath their protective elytra, what most people call their shell. The two elytra (singular elytron) are separated and lifted when the insect flies off.

Ecological Role   A gardener's friends, adults and larvae eat large numbers of aphids and other garden pests.

Plate 5

From top: COMMON EASTERN FIREFLY & LADY BEETLE

Honeybee: *Apis mellifera*

Description   A familiar sight on lawn-growing white clover and flowering garden plants, the three-fourth-inch-long honeybee is dull orange with black abdominal bands and transparent wings held close to the fuzzy body. The legs and head are blackish. Honeybee colonies are famous for the hive's social division of labor. Tens of thousands of sterile female workers provide pollen and nectar, tend the queen and young, and build and maintain the hive. The colony cannot survive without its queen. She is the egg-laying powerhouse. (In a hive, a few workers may lay eggs, but this number is trivial.) When ready to mate, the queen flies out of the hive and mates with a number of drones, or fertile males. Within the hive, a small number of drones develop from unfertilized eggs. Once grown, they leave to mate elsewhere with virgin queens ready to start new hives. Colonies last several years.

Ecological Role   In their quest for nectar, honeybees unwittingly pollinate many wild and commercial plants. Honeybee populations are in decline, plagued by colony collapse disorder, a threat brought on by varroa mite infestations. This has both ecological and commercial implications because honeybee pollination is crucial for commercial operations for apples, peaches, soybeans, pears, pumpkins, cucumbers, cherries, and other crops. Introduced from Europe as far back as colonial times, honeybees are raised for honey, wax, and other products. They also nest in the wild. You can look inside an active honeybee colony at the O. Orkin Insect Zoo, on the second floor of the National Museum of Natural History on the National Mall.

Eastern Carpenter Bee: *Xylocopa virginica*

Description   This round-bodied, 1-inch-long bee resembles a bumble bee with a hairless, or nearly hairless, black abdomen.

Ecological Role   Females dig into dead wood to lay their eggs and may be a nuisance when they excavate in and constantly fly around fences or wooden play gyms. These insects are not aggressive, however, and only sting when grabbed. This native bee is an important pollinator.

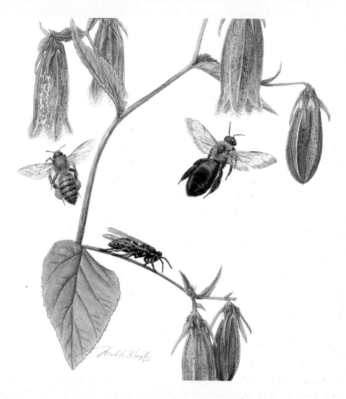

Eastern Yellow Jacket: *Vespula maculifrons*

Description   A bright yellow and black social wasp that grows to about half-inch long. Eastern yellow jackets build round underground nests in abandoned rodent burrows, sometimes in walls, under garden ties, or in rotting stumps. Colonies often contain thousands of workers. When a nest is disturbed, dozens of yellow jackets pour out and can inflict multiple burning stings. Eastern yellow jackets are active spring to fall. Yellow jackets are more aggressive than honeybees. Unlike bees, they can sting repeatedly.

Ecological Role   Eastern yellow jackets feed on plant nectar but also hunt other invertebrates including spiders and various insects. They are attracted to soft drinks, candy, and food, so they become a nuisance at tourist sites near outdoor food concessions and trash cans, particularly in late summer and early fall.

Eastern Cicada Killer: *Sphecius speciosus*

ETYMOLOGY

Sphecius speciosus: Latin for showy

Description   One of the largest North American wasp species, the eastern cicada killer hunts alone in sunny gardens, parks, lawns, and at the forest edge. Growing up to 2 inches long, with a bold-patterned yellow and black abdomen and orange legs, this is a gaudy insect but not a dangerous one. Unlike smaller, much more aggressive eastern yellow jackets, cicada killers rarely sting people. Only if someone steps on or handles a female will she sting. Males cannot sting.

Common Locations   Eastern cicada killers may be seen in sunny open areas with exposed patches of soil. Likely spots include the National Mall, the Smithsonian's National Zoo, Roosevelt Island, and around office buildings with sunny lawns. (They can be a regular sight around the State Department, for example.)

Notes of Interest   When a female cicada killer locates a cicada, she stings it, paralyzing the large insect. Then she drags or flies it back to her burrow, where she has laid her eggs in individual chambers. In each chamber, she places one to three cicadas. When the larvae emerge from the eggs, they feed on their allotted cicadas.

Ecological Role   Cicada killers feed on plant nectar, but females hunt cicadas to feed their young.

KEY POINTS

- A cicada killer's burrow may measure up to 4 feet long.
- The cicada killer is one of the largest North American wasps.
- Males cannot sting and females are not aggressive to humans.

Plate 7

EASTERN CICADA KILLER

Annual Cicada: *Tibicen* sp.

Description This large, noisy insect produces one of Washington, D.C.'s most familiar summer sounds—a droning racket that builds to a crescendo then fizzles, only to repeat again. Another name, the dog-day cicada, refers to the hottest summer days, the peak time for males calling to prospective females. White below and green above, with transparent wings, adult annual cicadas are often found lying on the sidewalk or street. When picked up, they often fly off with a noisy, startling buzz. The *periodical cicada*, next set to appear in Washington, D.C. in 2021, is black with orange-red eyes, and yellow-veined wings. In the city, they emerge between late April and early May every 17 years. Unlike the annual cicadas, they are usually gone before summer.

Ecological Role Annual cicadas are an important food for birds during nesting season and for some small mammals. Female cicada killers capture them to feed their young. Cicada nymphs grow up underground, feeding on tree roots. They take three years to mature and emerge. When they leave the soil, they climb up a tree trunk or other vertical surface. Once in a fixed position, they emerge through a slit in the back of their nymphal cuticle. These parchment-like, orangey skins linger for many weeks, still hooked where the cicadas left them.

Asian Tiger Mosquito: *Aedes albopictus*

Description The bold white banding on the otherwise black body and legs gives this biting insect its name. Asian tiger mosquitoes are active all day long. They look unassuming, measuring only up to one-fourth inch long, but females deliver a painful, then itchy sting. The Asian tiger mosquito is a recent arrival, first appearing in the United States in Houston, Texas, in 1985, likely in used tires imported from Japan. By 1987, some were found in Baltimore. Today, this is one of the area's most abundant mosquitoes. Females lay their eggs just above standing water level in rain gutters, empty flower pots, buckets, baby pools, pet water bowls, and watering cans. Very little water is needed for successful hatching of the wriggly aquatic larvae.

Ecological Role Asian tiger mosquitoes and other mosquitoes are an important protein source for swallows, spiders, and other insect-eating animals. From late spring through early fall, especially

Plate 8

From top: ANNUAL CICADA & ASIAN TIGER MOSQUITO

on muggy days when the air is still, this insect can put a damper on
picnic plans. Only females sting. They need blood meals to nourish
their eggs. Laboratory and field testing implicates this mosquito as a
potential or proven vector for the following diseases: dengue fever
(outside the United States), dog heartworm, eastern equine encepha-
litis, and West Nile virus.

Common Green Darner: *Anax junius*

Description   One of the largest dragonflies found in and around the city, the common green darner grows to 3 inches long, with a wingspan of up to four and a half inches. The head is large, the thorax is bright green, and the long abdomen is striped blue in males and green in females.

Ecological Role   Almost wherever there is water, darners and other dragonflies cruise in search of mosquitoes, midges, and other small flying insects. Often, they sit on the blades of grasses or on twigs. Because of its large size, the common green darner also takes larger prey, including butterflies, bees, and wasps. The larvae, also called naiads, capture prey, including small fish and tadpoles, in the muddy bottoms of ponds and in other slow to still waters.

Eastern Forktail: *Ischnura verticalis*

Description   This aerial hunter is a damselfly. Damselflies are closely related to dragonflies but tend to be much daintier and more slender. Damselflies hold their wings together above the body while at rest. Dragonflies' wings stay spread while at rest.

Ecological Role   Like their larger relatives, damselflies eat insects and other small invertebrates.

Eastern Amberwing: *Perithemis tenera*

Description   Unlike the large darner, the eastern amberwing is a small but stocky dragonfly. The male has a brown thorax and all-orange wings. The female has clear wings splotched with dark brown.

Ecological Role   Nymph and adult amberwings eat a wide variety of smaller insects.

Plate 9
Clockwise from top: COMMON GREEN DARNER,
EASTERN FORKTAIL, JJEASTERN AMBERWING

Polyphemus Moth: *Antheraea polyphemus*

Description   Named for the famed Cyclops in Homer's *The Odyssey*, this large moth makes a big impression with the false eye pattern on its burnt brown to yellow-brown wings. Each hindwing is adorned with a large "eye" outlined in black with purplish blue and a yellow center, while each forewing has a much smaller eye. The polyphemus moth has a 4- to 6-inch wingspan. Males have larger, more feathery antennae than females. The polyphemus moth, like most moths, is active at night and during warm months and sometimes shows up at outdoor lights. During the day, it may be found resting on leaves in forested parks. The caterpillar is striking as well: lime green with spots of red and silver, as well as yellow banding.

Ecological Role   The chunky caterpillar feeds on the leaves of oaks, maples, and other deciduous trees. Eastern screech-owls are among this insect's nocturnal hunters.

Eastern Tent Moth and Eastern Tent Caterpillar: *Malacosoma americanum*

Description   A soft brown, striped moth with a wingspan no greater than 1.5 inches, the eastern tent moth is best known in its caterpillar phase. Each April, just as leaves start to emerge, gauzy white nests appear in the crotches of cherry, crabapple, apple, and other trees in the rose family. These spring nests are made by silk-spinning eastern tent caterpillars that hatch from egg masses laid by female moths the previous summer. A white and gold stripe runs down the middle of the fuzzy caterpillar's back and dark spots mark the sides. The hundreds of caterpillars stay in the nest at night, leaving it during the day to feed. The caterpillars grow up to 2 inches long. By late spring, they disperse and spin white cocoons in which they metamorphose into moths.

Eastern tent moths can be seen around outdoor lights during the summer months.

Ecological Role   Voracious leaf-eating caterpillars often thin tree canopies, sometimes even defoliating host trees. This may make the trees more vulnerable to pests and diseases. Eastern tent caterpillars provide important springtime food for cuckoos, orioles, and many other birds. Blue-gray gnatcatchers sometimes pick the silk from eastern tent caterpillar nests for use in constructing their own. Birds and bats also feed on the moths.

Plate 10

From top: POLYPHEMUS MOTH & EASTERN TENT CATERPILLAR

Cabbage White Butterfly: *Pieris rapae*

Description   A European import in the nineteenth century, the cabbage white is now one of the most abundant North American butterflies. It's hard to overlook this insect's swerving, fluttery flight and its bright, tissue-like color. A smallish butterfly, the cabbage white has a wingspan of at most 2 inches. Active early spring to fall, a female usually produces several generations of young a year. Ash gray tips adorn each upper forewing. Distinguish genders by counting spots on the upper side of the forewing: males have one gray spot, females have two. Below, the cabbage white's wings are plain white and pale yellow.

Ecological Role   The cabbage white pollinates and feeds on the nectar of some of the city's most abundant plants, including dandelion and red clover. The bright-green caterpillars, called cabbage worms, feed on cabbages and nasturtiums, as well as crops and weeds in the mustard family.

Mourning Cloak Butterfly: *Nymphalis antiopa*

Description   With a wingspan up to 4 inches, the mourning cloak is larger than the cabbage white but a bit smaller than an eastern tiger swallowtail. Above, the wings are rust brown fringed with buttery yellow. The yellow fringe is edged by blue spots inset on the brown wings. Long-lived for a butterfly, the mourning cloak emerges in summer, overwinters, and mates in spring before dying. The mourning cloak is one of the first butterflies active in early spring. It is sometimes seen during winter warm spells. (The same holds true for question mark and comma butterflies.)

Ecological Role   The black, leggy caterpillars feed on the leaves of elms, willows, and other trees. Adults primarily eat tree sap and the juices of decomposing fruit, sometimes going to flowers for nectar.

Plate 11
From top: CABBAGE WHITE BUTTERFLY
& MOURNING CLOAK BUTTERFLY

Eastern Tiger Swallowtail: *Papilio glaucus*

Description   Very large and yellow with black tiger stripes and at least some blue on the hindwing, the eastern tiger swallowtail is probably the city's showiest common butterfly. Some females are blackish, but they always show at least some of the striped wing pattern. From late spring to fall, the eastern tiger swallowtail is a familiar sight in backyards and parks, where it feeds on a wide variety of flowers, often allowing close approach. With a wingspan of up to 4.5 inches, it is hard to miss.

Ecological Role   The bright green, bulgy caterpillar has a fake eye and a yellow collar. It feeds on tuliptree, willow, cherry, and other trees' leaves.

Spicebush Swallowtail: *Papilio troilus*

Description   This large, dark swallowtail frequents forest edge and backyards near parks. Males have a greenish sheen to the hindwings; females have a bluish sheen. On the underside of the hindwing, this butterfly has two distinctive bands of orange spots.

The somewhat similar black swallowtail is found in Washington, D.C., gardens and parks. The male has a yellow stripe across the lower hindwings.

Ecological Role   Spicebush swallowtail caterpillars feed on spicebush and sassafras, two plants abundant in Rock Creek Park and at the C&O Canal, among other places.

Plate 12
From top: EASTERN TIGER SWALLOWTAIL
& SPICEBUSH SWALLOWTAIL

Monarch: *Danaus plexippus*

ETYMOLOGY
Danaus: In Greek mythology, the king of Argos, who ordered his daughters to slay their husbands; *plexippus*: refers to one of the slain husbands

Description   Bright burnt orange outlined and veined in black, the monarch is one of the most familiar butterflies. The black fringing the wing is dotted with white, as is the body. Wingspan may reach 4 inches.

Common Locations   From summer to fall, the monarch brightens the city's parks and backyards, seeking nectar from milk-weed, joe-pye-weed, asters, purple coneflowers, black-eyed Susan, and other garden and field flowers. From September into October, monarch migration reaches its peak, and these orange beauties can be seen flying with a purpose over the city.

Notes of Interest   The *viceroy* mimics the monarch but can be identified by the black line that runs through the middle of its hindwings.

Ecological Role   The gaudy black-, yellow-, and white-ringed caterpillars grow up to 2 inches long. After they hatch, they start feeding on their host milkweed plants, which contain chemicals that make them and adults distasteful to predators.

KEY POINTS

- Monarchs from the eastern and central United States and Canada winter by the millions in a few spots in central Mexico. The same butterflies that winter in Mexico, however, do not make it back there the next year. After their one winter in Mexico, they start to move north in early spring, mating, laying eggs, and dying after finding milkweed patches in the East along the way. Three or four "generations" are produced as monarchs keep moving northward. The last is the late-summer generation, which somehow manages to migrate south to the traditional Mexico wintering grounds. A similar phenomenon occurs in western populations, which winter in parts of California. In the southern part of their range, monarch populations do not migrate.

- Females lay their eggs on milkweed plants. When they hatch, caterpillars feed on their host plant, growing quickly. When it is time to metamorphose, caterpillars wriggle into a green covering they excrete. This substance quickly hardens around

Plate 13
MONARCH

them, becoming the tent-shaped, bright-green chrysalis from which the butterflies emerge.

- Aside from those mentioned on previous pages, other common Washington, D.C., butterfly species include the question mark, the eastern comma, American lady, painted lady, clouded sulphur, silver-spotted skipper, sachem, gray hairstreak, spring azure, and eastern tailed-blue.

Crustaceans

Common Pillbug: *Armadillidium vulgare*

ETYMOLOGY
Armadillidium: small armadillo; *vulgare*: common

Rough Sowbug: *Porcellio scaber*

ETYMOLOGY
Porcellio: little pig; *scaber*: scaly

Description   While most people picture crabs, lobsters, and crayfish when they hear the word "crustacean," isopods make up another crustacean group, most of which live in marine environments. Some, however, live in freshwater habitats, while others live on land, including the slate-gray pillbugs and the usually slate-blue sowbugs, both of which are also called woodlice. These flattened, smooth-armored creatures, the size of an oblong pill, have seven overlapping, arched body segments with seven pairs of legs. Out the back end protrude two uropods, structures that collect moisture and dispose of waste. Pillbugs and sowbugs breathe through abdominal appendages, which they must always keep moist with droplets of dew or rain. Pillbugs are smooth and their sides are more or less even. Sowbugs have a rougher texture and lobed edges to their sides, with more prominent uropods.

Common Locations   Gardens and forests provide these interesting, abundant soil-living animals with the organic matter and humidity they need to survive. Pillbugs and sowbugs are active at night, seeking shelter during the day under flowerpots, logs, and other cover.

Notes of Interest   The Rock Creek Valley supports two tiny crustaceans not found anywhere else: the *Kenk's* and *Hay's Spring amphipods*.

Ecological Role   While feeding and burrowing, pillbugs and sowbugs aerate the soil and help produce humus. They eat algae, fungi, moss, and decaying plant and animal matter.

KEY POINTS

- These creatures are also called potato bugs and roly-polies because pillbugs roll up into a protective, armadillo-like ball when faced with danger. (Sowbugs don't curl up, but run from danger.)

INVERTEBRATES: CRUSTACEANS

162

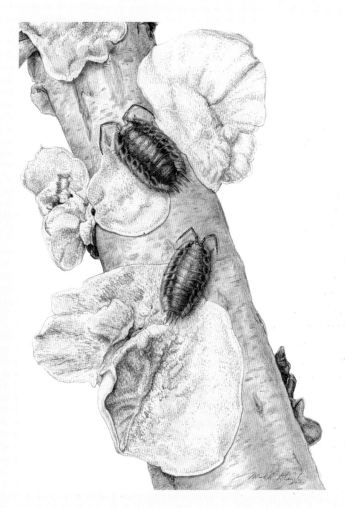

Plate 14
ROUGH SOWBUGS

- Sometimes you may find a strikingly bright-blue pillbug or sowbug. These bold-colored individuals are infected with a virus that strikes a variety of invertebrates.
- Although land-based creatures, pillbugs and sowbugs are still moisture dependent and brood their young in fluid-filled pouches called *marsupia*.

Crayfish: *Orconectes* spp.

ETYMOLOGY

Orconectes: *orco* refers to Orcus, a god of the underworld in Roman mythology; *nectes*: swimmer

Description   Mud-colored, armored "mini lobsters" that are in fact freshwater relatives of lobsters. While foraging, crayfish slowly walk forward. When they try to escape predators, they shoot backward, clenching their strong abdominal muscles. Crayfish have two pairs of antennae, a long one and a short, double-branched pair. They have four pairs of walking legs and the large chelipeds, or claws, which they use to capture prey and to defend themselves.

Common Locations   Crayfish are found in the city's rivers and streams, including Rock Creek. They are found along the C&O Canal as well. They hide in crevices and under loose rocks at stream bottoms, and some burrow into damp stream banks. During prolonged dry spells, they concentrate in areas with remaining water. When it rains, they can move short distances over land from one wet spot to another.

Notes of Interest   Aquaculturists and anglers dumping bait likely introduced the nonnative crayfish that now inhabit local streams. In Washington, D.C., six native species occur, but two introduced species, the virile and the red swamp crayfish, may be having a negative impact on these. The introduced rusty crayfish is also making inroads and may soon arrive to the city via the Potomac River.

Ecological Role   Crayfish are both predators and scavengers. They capture small animals but also feed on decaying organic matter they find on the stream bottom. Raccoons, night-herons, largemouth bass, and channel catfish are among their predators.

KEY POINTS

- Crayfish eggs develop and hatch below the female's abdomen. The young, which look like tiny adults, remain under her tail for a while before striking out on their own.

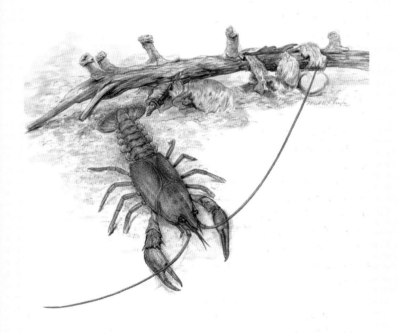

Plate 15
CRAYFISH

Common Carp: *Cyprinus carpio*

ETYMOLOGY

Cyprinus: Greek for carp; *carpio*: Latin for carp

Description   A large, sluggish fish seen in the city's murky shallows. These shiny, large-scaled fish are olive colored above, fading to whitish or yellowish below. From each side of the upper jaw protrude two barbels, or "whiskers." The rear barbel is much longer than the front one.

Size   The common carp grows up to 4 feet long and can weigh close to 100 pounds.

Common Locations   Common carp inhabit the Tidal Basin, Washington Channel, parts of the C&O Canal, and the Potomac and Anacostia Rivers.

Notes of Interest   Large carp put up a good fight, and angling for carp is a challenging fishing specialty that has grown popular.

Ecological Role   "Water hog" might be a good name for this adaptable introduced fish, which is native to Eurasia. The common carp thrives in almost any water quality, living in lakes and rivers. In some areas, these fish flourish in waterways polluted by farm runoff or sewage. As they feed, carp root around, stirring up plants and churning up mud. This activity often damages native fish breeding areas. The increased turbidity reduces sunlight that reaches remaining aquatic plants and makes finding prey more difficult for fish such as bass and sunfish. Common carp eat plant matter, invertebrates, and fish fry and eggs. Anglers can catch them using dough balls, corn, imitation berries, or worms as bait.

KEY POINTS

- Washington, D.C., waters are home to both introduced and native fishes. Introduced species include the common carp, smallmouth and largemouth bass, bluegill, and channel catfish and blue catfish. The northern snakehead, first found in the Potomac in 2004, is a long predatory fish from Asia. It was caught recently in the Tidal Basin and elsewhere. Native fish found in the Potomac include black and white crappies, white and yellow perch, American shad, pumpkinseed, striped bass, and American eel.

Plate 16

COMMON CARP

Channel Catfish: *Ictalurus punctatus*

ETYMOLOGY

Ictalurus: fish cat; *punctatus*: spotted

Description   Catfish are distinct, with large eyes, wide mouths, and whisker-like barbels. Unlike bass, carp, and sunfish, channel catfish, and the somewhat similar blue catfish, lack scales. Channel catfish are long and slender, with dark grayish or greenish backs, fading to a lighter shade on the sides. A light scattering of dark spots fleck the sides of young adult and juvenile channel catfish, a field mark useful in distinguishing this from other species. Also, the anal or lower rear fin is rounded, unlike the blue catfish's straight-edge anal fin. Catfish appear much more whiskered than carp. While common carp have two pairs of barbels on the upper lip, the channel catfish has a long pair sweeping back from upper jaw to below its pectoral fins, two barbel pairs on the chin, and a pair atop the nose. The caudal, or tail, fin is deeply notched.

Size   A record channel catfish reached almost 4 feet long and weighed up to 58 pounds. Local fish are far smaller. The introduced blue catfish is on the increase in the tidal Potomac. It grows much larger, to five and a half feet long and well over 100 pounds.

Common Locations   Channel catfish favor rivers and large creeks with slow to moderate current. They live in the Potomac and Anacostia Rivers, their larger tributaries, and in the Tidal Basin.

Notes of Interest   Native to North American river systems west of the D.C. area, the channel catfish was introduced and now flourishes in and around the city's rivers and larger creeks. Thanks to aquaculturists and anglers, channel catfish also found their way to many other parts of North America, as well as Russia, Brazil, China, and other countries.

Ecological Role   Depending on their size, age, and location, channel catfish will eat a wide variety of both animals and plants. They tend to feed near the bottom but will rise if opportunity knocks. Prey includes many types of insect, other fish, crayfish, snails, and, rarely, birds.

KEY POINTS

- Channel catfish don't have scales, and they have four sets of barbels for sensing prospective food items.
- They eat a wide variety of plants and animals, usually seeking food near the bottom.

Plate 17

CHANNEL CATFISH

- Channel catfish have rigid spines in their dorsal and pectoral fins
  that can cut anglers who don't know how to handle them.

Bluegill: *Lepomis macrochirus*

Description   The bluegill's body is laterally compressed, appearing as if its sides were squished in. When facing head on, it looks like a vertical pancake with eyes. Bluegill is distinguished from the pumpkinseed and other sunfish by the large black spot adorning the rear of its dorsal fin. Male and female bluegill also have larger black "ears" extending from the back of the gill covers, or opercula, than do pumpkinseeds. The female is silvery above, with darkish vertical bars. The male also has barring but is greenish with orange on the breast. The fish's name comes from the pale-bluish coloration below the cheeks and along the "gill rakers."

Ecological Role   Streams, rivers, and lakes often abound with bluegills, which have been introduced for fishing. Bluegills are omnivores, eating a wide variety of invertebrates and tiny fish. In turn, they are eaten by largemouth bass, catfish, and other larger fish, as well as turtles, herons, and ospreys.

Common Locations   From Constitution Gardens to the rivers, this is a very common fish.

Notes of Interest   Bluegill often school in groups of 20 to 30. For nests, they sift away thin soils, creating shallow bowls. Males guard these nests.

Largemouth Bass: *Micropterus salmoides*

Description   This silvery fish makes an impression when hooked. It grows to about 2 feet long, with exceptional records over 3 feet and just over 20 pounds. Unlike bluegill, the body is elongated. The largemouth bass has a dorsal fin with two distinctive, almost separated parts. The first part is spiny and highest in the middle. The second part is soft, rounded, and more gently curved. The largemouth bass is white below and grayish above, with a broad blackish stripe along the sides. True to its name, it has a large, and longer, mouth than the smallmouth bass. The largemouth's upper jaw line extends backward well beyond the eye. The smallmouth bass is also caught in D.C. waters. This fish has greenish-yellow sides punctuated with distinct blotchy bars and a dorsal fin that does not appear divided, as in the largemouth.

Ecological Role   The largemouth bass is a predator of ponds, lakes, and mud-bottomed streams. Its varied diet includes aquatic invertebrates, other fish, amphibians, birds, and mice.

Plate 18

From top: BLUEGILL & LARGEMOUTH BASS

Common Locations   The Washington Channel, which visitors to East Potomac Park face on the east side, is a well-known spot for bass and carp fishing. The largemouth bass is common in the Potomac, at the Tidal Basin, deeper parts of Rock Creek, and elsewhere.

Notes of Interest   A very popular game fish that has been introduced to many parts of the continent, including the D.C. area.

Redback Salamander: *Plethodon cinereus*

ETYMOLOGY

Plethodon: refers to full of teeth; *cinereus*: ash-colored, descriptive of leadback color phase

Description   The small redback salamander can be an abundant amphibian below logs and other debris in undisturbed, shady woodlands. The copper-colored eyes have a marble-like appearance, and the snout is rounded and short. There are two common color phases—the redback and the leadback. Redbacks have a broad reddish or orangish stripe down the back and the top of the tail. Leadbacks have a broad gray stripe matching their gray sides. All redback salamanders have a salt-and-pepper pattern below, with equal parts black and white stippling. Although often confused with lizards, salamanders are not related. Lizards are reptiles; salamanders (including newts) are amphibians. Salamanders lack the scaled skin, claws, and ear openings of lizards. Redback salamanders, like most salamanders, move around and hunt at night. The five-lined skink, the most likely lizard to be seen in Washington, D.C., is active during the day.

Size   Redback salamanders grow up to 4 inches long; most individuals found are smaller than this.

Common Locations   Rock Creek Park and the adjacent fringe of the National Zoo are ideal sites for redback salamanders, where there is ample undergrowth and large stones, decaying logs, thick leaf litter, even trash—under which these animals seek refuge during the day. On warm, rainy nights they are most active and can be found by walking the woods with a flashlight or headlamp. Check with park staff to see when and where nighttime salamander watching is permitted. Most parks close after dark.

Notes of Interest   Redback salamanders lay their eggs in May and June. Young develop within the egg, skipping the aquatic larval stage found in many other salamanders. Tiny young salamanders may be found in the fall. Somewhat similar in appearance to the redback, the northern two-lined salamander lives in streams in Rock Creek Park and Glover-Archbold Park, among other places.

Ecological Role   The redback salamander hunts small insects, spiders, earthworms, and other invertebrates. In turn, bullfrogs, snakes, blue jays, raccoons, and other predators eat them.

Plate 19

REDBACK SALAMANDER

KEY POINTS

- While newts and many other salamanders frequent watery habitats, the redback salamander is a woodland creature, laying its eggs under or in rotting logs, moss, and other moist situations. It breathes through its skin and mouth and does not have lungs.
- Dryness is the salamander's mortal enemy: During rainless, hot weather, redback salamanders retreat to crevices, dig beneath the soil, and may go into a hibernation-like dormant state.
- The redback salamander is the most widely distributed and often the most common salamander in the region.

American Toad: *Bufo americanus*

ETYMOLOGY

Bufo: toad; *americanus*: of America

Description   Unlike frogs, the American toad's body is dry and bumpy. Body color varies from plain brown to brick reddish, with small light-colored patches. Sometimes a pale line runs down the spine. On the back, the warts (actually glands) usually occur one to each dark spot. Those on the thighs usually have a spine. Below, American toads are usually black spotted.

Size   American toads grow up to 4 inches from snout to the end of body. Females are usually larger than males.

Common Locations   During the early spring breeding season, American toads can be found around slow-moving streams, in watery ditches, and in other shallow pools where they breed. At this time, you can hear them along Rock Creek and the C&O Canal and at the Kenilworth Aquatic Gardens. From late spring to early fall, they lay low in forested areas, in burrows, or under debris, coming out at night to hunt.

Notes of Interest   During warm days of late winter and early spring, listen for the American toad's drawn-out trill. One trill can last up to 30 seconds. The similar Fowler's toad has a shorter, sheep-like call and usually calls and breeds later in spring. Identifying Fowler's from American toads is tricky for beginners. Fowler's is more often found in sandy soils.

Ecological Role   A nocturnal hunter of insects and other invertebrates, the American toad is hunted by snakes, raccoons, and other predators.

KEY POINTS

- You will not get warts from handling a toad, although it might be stressful for the toad, and its gland secretions might irritate your nose and eyes if you rub them before washing your hands.
- Toad coloration can vary, depending on weather conditions and the toad's mood.
- As in most frogs and toads during breeding periods, male American toads piggyback onto females in an act called amplexus, fertilizing egg strings as they are deposited in the water.

Plate 20

AMERICAN TOAD

Bullfrog: *Rana catesbeiana*

ETYMOLOGY

Rana: a frog; *catesbeiana*: in honor of Mark Catesby, an English naturalist who surveyed wildlife in the 1700s in what would later become the southeastern states

Description   North America's largest frog. The bullfrog's body color is usually green but may be brownish. Bullfrogs are whitish below, but the male has a yellow throat during breeding season.

Size   The head-body length of this frog averages 3.5 to 5 inches. The record, though, is 8 inches.

Common Locations   Bullfrogs tend to live in larger water bodies, such as marshes, lakes, and the calmer parts of streams. But they inhabit small ponds as well, including those at the National Zoo. Also, watch for bullfrogs at Kenilworth Aquatic Gardens, the C&O Canal National Historical Park, and Rock Creek Park.

Notes of Interest   Another common frog, the green frog, looks similar but is much smaller, with noticeable ridges running down the sides of its back. Its call is different too (see below).

Ecological Role   Many of us think frogs are just hapless prey, but the bullfrog is a voracious predator in its own right, catching pretty much whatever comes its way, as long as it is alive and will fit in its generous-sized mouth. Prey may include small snakes, ducklings, hummingbirds, chipping sparrows, and other small birds, small turtles, other frogs and tadpoles, salamanders, fish, small rodents, and insects, spiders, and other invertebrates. Raccoons, herons, water snakes, snapping turtles, and largemouth bass catch and eat bullfrogs and their large tadpoles.

KEY POINTS

- Bullfrogs breed from April well into the summer.
- Listen for the males' low "jug-o-rum" call, which differs from the banjo-like twang of the smaller green frog.
- Females lay 10,000 to almost 30,000 eggs among wetland vegetation, and tadpoles may take up to two years to metamorphose into frogs.

Plate 21
BULLFROG

Spring Peeper: *Pseudacris crucifer*

ETYMOLOGY

Pseudacris: Greek for false locust (for sounds); *crucifer*: Latin for cross-bearer for back pattern

Description   Though often heard, this small frog is rarely seen. The body color may be tan to gray, but all individuals are marked with a distinctive X-like cross on the back. A dark bar runs between the eyes. Spring peepers are white to cream below. During breeding season, males can be told from females by their dark throats. Peepers' high-pitched, somewhat cricket-like peeps can be heard at a time when few or no insects are sounding off—on warmer late winter and early spring evenings. Choruses of these frogs take place along wetland edges and in swamps and even in the city can be heard from at least several hundred yards away.

Size   Spring peepers rarely exceed 1 inch in head and body length.

Common Locations   Listen for spring peeper choruses in Rock Creek Park, Kenilworth Aquatic Gardens, the C&O Canal National Historical Park, and other places with low, wet areas and forest. Armed with patience and a good flashlight, the intrepid night observer may spot a few of these tiny frogs at their breeding pools. (Check with park staff before making any night wanderings.)

Notes of Interest   Spring peepers spend much of their year in the forest but congregate in wetlands for breeding.

Ecological Role   Spring peepers catch beetles, flies, ants, spiders, and other small invertebrates, while keeping an eye out for their own predators, including bullfrogs, water snakes, and herons.

KEY POINTS

· If night temperatures reach the 50 to 60 degrees Fahrenheit in January or February, spring peepers start to call. Peak time, however, is from late March into early April.

· Spring peepers sometimes call on overcast days, and a few may call intermittently in autumn.

· Scientists have puzzled over how to classify the spring peeper. It has characteristics recalling two frog groupings, the chorus frogs and the treefrogs. Most recently, spring peepers have been placed with the chorus frogs.

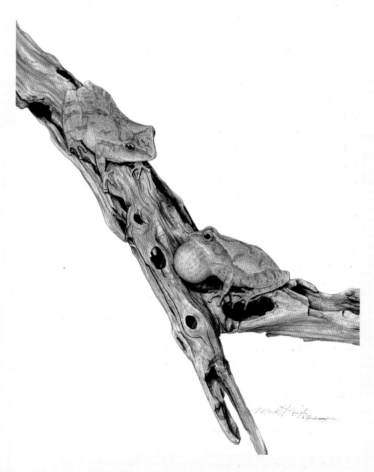

Plate 22

SPRING PEEPER

Snapping Turtle: *Chelydra serpentina*

ETYMOLOGY

Chelydra: tortoise; *serpentina*: snake like

Description   A primitive-looking dark brown and gnarled turtle with a long bumpy tail, elephantine legs tipped with long claws, long neck and bullish head, and undersized looking shell.

Size   Often, a grown snapping turtle's carapace, or top shell, exceeds a foot in length, and the tail is about as long as the carapace. Adults normally weigh between 10 and 35 pounds.

Common Locations   Snapping turtles spend a lot of time underwater and undercover, but they inhabit most permanent freshwater areas in and around the city. They may be seen at Kenilworth Aquatic Gardens, C&O Canal National Historical Park, Roosevelt Island, and many other areas. Shallow waters, such as a drying section of the C&O Canal or a low open area of marsh may reveal one slowly moving across, just below the water's surface.

Notes of Interest   Snapping turtles often bury themselves in mud. While swimming, they usually keep mostly below the water's surface, nose and eyes only visible as a bump in the water.

Ecological Role   Snapping turtles eat beetles, fish, reptiles, ducklings, and other small animals, but they also eat carrion and plant matter. Raccoons, skunks, and foxes dig up their nests to feast on the eggs. Hatchlings fall to crows, herons, bullfrogs, largemouth bass, and other predators.

KEY POINTS

- Temperature likely determines the sex of snapping turtle young. Within a nest, warmer eggs usually produce females, cooler eggs are usually males.
- Unlike painted turtles, snapping turtles seldom bask at the water's edge. In spring, females haul up on dry land to dig nests and lay their eggs.
- Despite popular belief, snapping turtles pose little danger to waders, even pulling in their heads when accidentally stepped on. However, if threatened on land, they will lash out.

Plate 23

SNAPPING TURTLE

Eastern Painted Turtle: *Chrysemys picta*

ETYMOLOGY

Chrysemys: golden freshwater tortoise; *picta*: painted

Description　The eastern painted turtle is the size and shape most people imagine for a typical turtle. The top shell, or carapace, looks blackish when seen from a distance but is actually dark brown or dark olive with pale borders between the scutes, or plates, of the shell. The margins of the carapace are often scalloped with red and black. Adults often have two pale spots behind each eye, as though someone dabbed yellow paint there.

Size　The carapace rarely measures longer than 7 inches.

Common Locations　This is the most common basking turtle in the area. On warm days, it hauls itself out onto logs jutting out of the water or sits at the water's edge. The eastern painted turtle is frequently seen at the C&O Canal National Historical Park, Kenilworth Aquatic Gardens, the National Arboretum, Roosevelt Island, Rock Creek Park, and other sites.

Notes of Interest　Two other basking turtles commonly occur in the area: The larger redbelly turtle has a more domed carapace, often with red stripes, and the introduced red-eared slider is olive green with a red "ear" patch.

Ecological Role　Painted turtles are omnivores. Their diet may include beetles, fish, carrion, and aquatic plants and algae. Their eggs are eaten by raccoons, skunks, and foxes, among other mammals. Young are eaten by raccoons, snapping turtles, largemouth bass, herons, crows, and other predators.

KEY POINTS

- Although they are most active between March and October, painted turtles may be out basking on very warm winter days.
- During winter cold snaps, these turtles hibernate in water, in muskrat or beaver lodges, or under logs or other debris.

Plate 24

EASTERN PAINTED TURTLE

Northern Water Snake: *Nerodia sipedon*

ETYMOLOGY

Nerodia: moving through (reference to water); *sipedon*: perhaps refers to a snake with a bite that causes mortification; origin is unclear

Description   A fairly large, thick-bodied snake usually seen in or near water. Depending on age, stage of molt, or individual coloration, its back markings may appear clearly banded to uniform dark. Usually, some banding is visible. Young snakes are boldly marked in cross bands of gray and black.

Size   The northern water snake rarely grows longer than 40 inches, but the record was an individual that reached 55 inches.

Common Locations   Most unmanicured wetlands in the city support northern water snakes. You may see them at Roosevelt Island, C&O Canal National Historical Park, Kenilworth Aquatic Gardens, Rock Creek Park, and other wet areas.

Notes of Interest   The northern water snake's thick body and markings sometimes mislead people into thinking it is a poisonous cottonmouth, a species not found in the city, with its closest populations in southern Virginia. The city's only poisonous snake, the northern copperhead, is not usually found near water and is only rarely reported in the city, though it is found in suburban parks just outside. No copperhead sightings have been confirmed in Rock Creek Park since the 1970s.

Ecological Role   Northern water snakes primarily hunt fish and amphibians, including young carp and goldfish, toads, and frogs. They in turn feed raccoons, largemouth bass, snapping turtles, and herons.

KEY POINTS

- Female northern water snakes usually grow longer than males.
- Although not poisonous, water snakes have long, curved teeth and can inflict a nasty bite if provoked. However, they normally are not aggressive.

Plate 25

NORTHERN WATER SNAKE

Black Rat Snake: *Elaphe obsoleta*

ETYMOLOGY

Elaphe: a kind of snake; *obsoleta*: worn out, referring to the loss of patterning on adults

Description The long, sleek black rat snake is one of the most common eastern snakes. Adults are uniform black above with black and white checkering below. The chin and throat are usually whitish. Young black rat snakes are crisply patterned above—gray with black or dark brown patches.

Size Black rat snakes are the largest snakes normally encountered in the city. The record length is 101 inches, but lengths between 40 and 70 inches are commonplace.

Common Locations Watch for black rat snakes in wooded parts of the Kenilworth Aquatic Gardens, Roosevelt Island, the C&O Canal Historical Park, Rock Creek Park, and other parks. They may be found near water, inside the forest, on the ground near trails or seen climbing the trunks of trees.

Notes of Interest Black rat snakes kill their larger prey by constriction. They swallow eggs, which break in their throats.

Ecological Role Black rat snakes primarily eat rodents, birds, and birds' eggs. Prey includes gray squirrels, eastern chipmunks, mice and voles, American robins, gray catbirds, and downy woodpeckers. Large owls, red-tailed and red-shouldered hawks, and cats are among their predators.

KEY POINTS

- Black rat snakes are active both day and night. Many are killed on roadways.
- Black rat snakes are harmless. Some are docile, but some will strike if harassed.
- In Washington, D.C., snakes usually mate from April through May or in some cases in early summer. Eggs hatch from July through September.

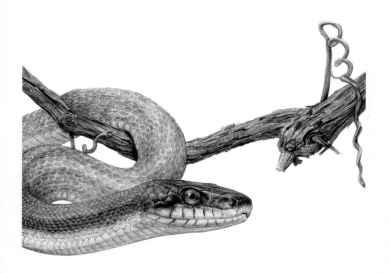

Plate 26

BLACK RAT SNAKE

Double-crested Cormorant: *Phalacrocorax auritus*

ETYMOLOGY

Phalacrocorax: bald crow or raven; *auritus*: eared

Description Adult double-crested cormorants are large, blackish, goose-sized waterbirds. While fishing, they dive beneath the water's surface, propelled downward by thrusts of their webbed feet. After feeding, double-crested cormorants usually leave the water to dry out on pilings, rocks, or sturdy waterside tree snags, holding their wings open to facilitate this process. In flight, cormorants are somewhat goose-like, even flying in V formation. But unlike geese, they are dark all over, with longer tails and slightly kinked necks. The only parts of an adult double-crested cormorant that aren't dark are the small orange throat poach and face and green eyes. Juveniles are two toned, with blackish bellies and pale-brown breasts and necks. The namesake twin head feathers can only be seen during the breeding season. Even then, they may appear frayed or irregular.

Size The size of a small goose, the double-crested cormorant reaches bill-to-tail lengths of on average 33 inches and has a wingspan of about 52 inches, or 4.3 feet.

Common Locations Double-crested cormorants frequent rivers, open marshes, bays, and other productive fishing grounds. They are a frequent sight on rocks, in water, or overhead in loose V-shaped formations. Often seen on the Potomac and Anacostia Rivers.

Notes of Interest In the city, peak migration periods for double-crested cormorants are from March to May and August to November, when the birds use the Potomac as a migration corridor. They are scarce from December through February.

Ecological Role Double-crested cormorants catch a wide variety of fish and crustaceans, amphibians, and other aquatic life.

KEY POINTS

- While swimming, cormorants often trick birders into thinking they are loons or other birds because of their low profile in the water, with only their periscope-like necks and heads and a little of their backs held above the water's surface.
- Cormorants are related to pelicans, anhingas, frigatebirds, tropicbirds, gannets, and boobies.
- The double-crested is the most widespread cormorant species in North America.

Plate 27

DOUBLE-CRESTED CORMORANT

Canada Goose: *Branta canadensis*

ETYMOLOGY

Branta: burnt, refers to burnt color; *canadensis*: of Canada, where most of the birds' nesting once occurred. (There are now also wide-spread breeding populations across much of the United States, including in Washington, D.C.)

Description   This bird is easy to identify without binoculars. A noisy, large goose with a softly hued brown body and long jet-black neck and head accented with white cheeks. The gleaming white vent—the area between the belly and the tail—also stands out.

Size   At about 45 inches long, with a wingspan of 5 feet, the Canada goose is the largest waterfowl species commonly seen in the city. Only swans are larger.

Common Locations   These noisy, large, gregarious birds are common all year in many of the city's watery areas, as well as in wide-open grassy spaces. You will likely see Canada geese at Roosevelt Island, Hains Point, the C&O Canal National Historical Park, Kenilworth Aquatic Gardens, and other locations.

Notes of Interest   In fall, winter, or early spring, a very small goose with identical markings and a stubby bill may turn up within a Canada goose flock. Once considered a race of Canada goose, the rare cackling goose was recently given separate species status. Snow geese, white with black wing feathers and pink bills, rarely show up in Washington, D.C., as do mute swans, very large white birds with elegant curved necks and knobby orange bills. Tundra swans—white with black bills—pass over in March and November.

Ecological Role   Primarily vegetarians, Canada geese snip grass shoots and stems and aquatic plants and guzzle seeds, grains, and berries. Insects and other small creatures sometimes wind up on the menu.

KEY POINTS

- The distinctive honks of Canada geese are one of the most familiar natural sounds in watery parts of Washington, D.C.
- The city's Canada geese may be joined in winter by some truly Canadian birds, but the majority of these winter on the nearby Delmarva Peninsula in Delaware, Maryland, and Virginia.

Plate 28

CANADA GOOSE

Mallard: *Anas platyrhynchos*

ETYMOLOGY

Anas refers to duck; *platyrhynchos*: broad-billed

Description The mallard is the most familiar and easily seen duck in the region. The adult male has a metallic-green head, a bright-yellow bill, and a white collar that separates the green head from the rust-colored chest. The body is lead gray and the tail is white surrounded by black, including distinctive curled feathers. The female is far less distinctive. Her best field marks are an orange bill with a dark blotch in the middle, and the dark line through the eye. Otherwise, her plumage is light brown accented with dark chevrons. Both male and female have bright orange legs and feet and on each wing a medium-blue speculum (or wing patch) bordered in white.

Size Almost 2 feet long, with domesticated varieties such as the farmyard "Pekin" duck reaching 2 feet. The mallard's wingspan averages 35 inches. The largest duck normally seen around Washington, unless a visitor comes across a released domestic moscovy, a strange-looking bird with a red, knobbed face and bill.

Common Locations Watch for mallards anywhere there is water: in the rivers, waterside parks, small ponds, rain-puddled playing fields, even reflecting pools on the National Mall.

Notes of Interest Many of the closely related American black ducks (Plate 29, *lower left*) seen in areas such as Roosevelt Island have at least some mallard blood. American black ducks have all dark bodies and tan heads. Birds with greenish crowns are hybrid males.

Ecological Role Mallards belong to the dabbling duck group, waterfowl that tip their bottoms up and submerge their heads to nibble aquatic plants and seek out various small aquatic animals such as snails and tadpoles. Mallards also waddle over land to graze on grasses and find seeds, grains, and insects.

KEY POINTS

- Abundant and approachable, mallards are the "pigeon of ducks." Like pigeons, they deserve a closer look because although tame and familiar, they are beautiful. Check out the drake's dazzling green head, for example.
- After leaving the nest, six or more ducklings follow their mother, sometimes making perilous trots down sidewalks, across grass islands, and across busy streets.

Plate 29

- Ducks have different vocalizations, varying by species. The mallard, ancestor of most domestic duck breeds, gives a "quack" familiar to most people. However, chicks peep.

Wood Duck: *Aix sponsa*

ETYMOLOGY

Aix refers to a type of waterfowl; *sponsa*: refers to "betrothed or wedded" due to the bird's dapper appearance

Description One of the world's most beautiful ducks, the wood duck is a wild bird that has adapted well to the forested city of Washington, D.C. Sometimes easy to view close up, the drake, or male, combines metallic green on the head and back with the colors of the local pro football team—an almost burgundy chest with goldish-yellow sides. The male's bill is bright red with dabs of black, while throat, cheek, and collar are white, contrasting the head's darkness. More somberly colored, the female is also distinctive: a horizontal teardrop of white surrounds her eye, tapering toward the back. Both sexes have a "hammer-headed" look. In flight, wood ducks' proportionately longer tails, bumpy heads, and short bills easily separate them from flying mallards.

Size Smaller and smaller billed than the mallard, the wood duck reaches 19 inches in length, with a 30-inch wingspan.

Common Locations Pairs or females with young frequent streams, rivers, and wetlands. In the city, these birds are often accustomed to the sight of joggers, cyclists, and walkers along Rock Creek and the Potomac at Roosevelt Island, at the C&O Canal, Kenilworth Aquatic Gardens, and in watery exhibits at the National Zoo. From November through early February, wood ducks are scarce in the city, but a few usually winter at the National Zoo's birdhouse ponds, where they act like part of the collection.

Notes of Interest Asia's mandarin duck is the wood duck's only close relative.

Ecological Role While they sometimes eat insects and small crustaceans, wood ducks mainly eat seeds. These include acorns and grains on land and aquatic plant seeds in the water.

KEY POINTS

- With a little practice, it's easy to identify wood ducks just by their distinctive voices. As they take off, these birds call a loud, distressed-sounding "Wwweeeeep!"
- Unlike many ducks, the well-named wood duck nests in cavities high up in hardwood trees, and in nest boxes made for their use. Tiny young wood ducks fledge by hopping out of the nest cavity and tumbling to the ground below.

Plate 30

WOOD DUCK

Ring-necked Duck: *Aythya collaris*

ETYMOLOGY

Aythya: a kind of waterbird; *collaris*: collared

Description   Ring-necked ducks resemble closely related scaup, which are often in larger bodies of water. The dark-headed male is separated from scaup by the white half-moon that separates his blackish chest from his pencil-gray sides. Also, his gray bill appears hand-painted with a white band and a black tip. The female is brown, with a grayish head, narrow white ring around the eye, and a similarly marked gray bill and a whitish wash where bill meets face.

Size   Much smaller than a mallard, the ring-necked duck averages 17 inches long, with a 2-foot wingspan.

Common Locations   Usually not close enough to spot easily without binoculars, the ring-necked duck is a fall, winter, and early spring visitor to such areas as Constitution Gardens, the C&O Canal, McMillan Reservoir, Kenilworth Aquatic Gardens, Roosevelt Island, and other spots along the Potomac and Anacostia Rivers. Favored habitats include rivers, ponds, and bays with woodland along the edge.

Notes of Interest   The ringneck is just one representative example of a duck that regularly winters in the Washington, D.C., area. There are many others, some mentioned in the More D.C. Birds section.

Ecological Role   An omnivore, the ring-necked duck feeds on a variety of submerged aquatic plants, as well as aquatic insects and mollusks.

KEY POINTS

- Unlike the mallard, which dabbles for its food in water, the ring-necked duck dives beneath the water's surface.
- The namesake collar on this duck is rarely seen in the field.
- The numbers of ring-necked ducks and other migratory waterfowl in Washington, D.C., vary widely from year to year. As the Great Lakes and other northern waterways freeze over, waterfowl drift southward. Waterfowl numbers also fluctuate depending on the breeding success during the previous summer in the Prairie Pothole region of the north-central states and Canada.

Plate 31
From bottom: RING-NECKED DUCK & HOODED MERGANSER

Hooded Merganser: *Lophodytes cucullatus*

ETYMOLOGY
Lophodytes: crested diver; *cucullatus*: hooded

Description For such a gaudy duck, the hooded merganser can be easily overlooked. It is small, and small groups often skirt the shoreline of rivers and lakes. Males can be mistaken at a distance for wood ducks, but up close there's no mistaking the black head punctuated by a large white wedge. When excited, the male raises his crest, changing his head shape from flattened to round, and the white wedge transforms into a large inverted comma. The male's black back and orange sides are separated from the white breast by two black bars. The eyes are piercing yellow. Females are dingy brown but with yellow eyes and a similar head shape to the male. Females of the larger red-breasted and common mergansers, which also occur in the city fall through early spring, have gray bodies and rusty heads.

Size Measuring 18 inches long, the hooded merganser is about as long as a crow.

Common Locations Hooded mergansers frequent freshwater lakes, marshes, and rivers. Watch for them along Potomac and Anacostia river sites, including the C&O Canal, Kenilworth Aquatic Gardens, and Roosevelt Island.

Notes of Interest The word *merganser* comes from the Latin for diving goose.

Ecological Role The hooded merganser catches small fish and crayfish and eats aquatic insects, along with a tadpole or two.

KEY POINTS

- Odd among ducks, mergansers' bills have saw-like, serrated edges that aid in capturing fish. These edges are modified lamellae, structures adapted in other duck species to strain out food items much like the baleen of whales.
- Unlike the two previously mentioned mergansers, which range across the Northern Hemisphere, the hooded merganser lives only in the United States and Canada, rarely straying across the border into Mexico.
- Hooded mergansers are most frequently encountered in D.C. from mid-November to mid-April.

Hooded Mergansers

Great Blue Heron: *Ardea herodias*

ETYMOLOGY

Ardea: Latin for heron; *herodias*: Greek for heron

Description   Great blue herons are the only very large, gray, long-legged wading birds in the city. People often mistake them for cranes or storks, birds rarely seen in the region. More gray blue than true blue, these birds fly over the city but are most often seen standing quietly at the water's edge, ready to strike at prey with their long orange bills. The long neck is tinged with a faint rose flush, and patches of red adorn the birds' shoulders and the "stockings" above the legs. Males and females have the same coloration. Adults sport black head plumes that contrast the otherwise whitish head. Young birds have blackish crowns.

Size   About 4 feet tall with a 6-foot wingspan, the great blue heron is one of the largest wild birds found in the city.

Common Locations   The shoreline and rocks along the Potomac River and marshes along the Anacostia usually host small numbers of great blue herons at any time of year. Check the C&O Canal, Roosevelt Island, and Kenilworth Aquatic Gardens, for example. During dry summer months, great blues congregate where there is water.

Notes of Interest   Great blue herons usually fly with their necks curled in, forming a bulge. The slow-flapping, long-legged great blue is easy to identify even high in the sky.

Ecological Role   An adaptable hunter of the shallows, the great blue heron frequents a variety of wetlands, including ponds. Fish are an important component of the diet, but frogs, snakes, mice, and even small birds are sometimes eaten.

KEY POINTS

- A few other herons can be seen in the city, but the only one that approaches the great blue's size and proportions is the all-white great egret, which appears along the river during migration and in summer.
- There are several great blue heron rookeries, or nesting areas, near the city, including one upriver along the Potomac in Maryland.
- The great blue heron sounds as though it is disgusted as it takes off from the water's edge, uttering an elongated *craaaaaahk*.

Plate 32

GREAT BLUE HERON

Black-crowned Night-heron: *Nycticorax nycticorax*

ETYMOLOGY

Nycticorax : night crow

Description Each spring, wild black-crowned night-herons continue a decades-old tradition and nest around the National Zoo's birdhouse. Among the capital's most unusual birds, adults are eye-catching, with the namesake black crown and back and a white head plume that sweeps back over it. Other field marks include the thick, black bill, fiery red eyes, and yellow legs. Silvery wings contrast the birds' white faces and underparts. By early June, brown, full-sized immature birds, their plumage streaked and spotted with white, clamber over branches, noisily begging food from their dapper parents.

Size About the size of a raven but with far longer legs, the black-crowned night-heron grows to about 25 inches long, with a 44-inch wingspan.

Common Locations The best place to see black-crowned night-herons close up is near their breeding colony at the National Zoo. The best time is from April well into the summer months. Most of the adults and immature birds vanish by the end of August. During nesting season, up to 200 nests festoon trees near the birdhouse, including those in the crane yards. Immature birds and adults frequent the pools in the zoo's enclosures, including the flamingo and stork pools behind the birdhouse and the kori bustard enclosure. But adults do most of their hunting along Rock Creek and the Potomac, where they are a regular sight along the rocky shoreline north of Roosevelt Island. They also roost in trees at the north end of East Potomac Park and at Ladybird Johnson Park.

Notes of Interest The black-crowned night-heron is found across North America and in many parts of South America, Europe, Asia, and Africa.

Ecological Role Black-crowned night-herons primarily eat fish but also take a variety of other creatures living in and around the water. These include frogs, snakes, rodents, crayfish, and carrion. They also snatch food put out for such National Zoo birds as the kori bustards.

KEY POINTS

- The black-crowned night-heron is a squat, chunky bird that usually appears short necked for a heron. It is medium-sized—

Plate 33
BLACK-CROWNED NIGHT-HERON

far smaller than the great blue heron and is far less likely to be seen out during the day away from its nesting area.

- Listen for the black-crowned night-heron's distinctive, choppy *quahk* as it flies overhead.
- A yellow-crowned night-heron rarely shows up at the zoo's nesting colony of black-crowned night-herons. There are a few spots just outside the city where this blue-necked, yellow-crowned bird nests.

Red-tailed Hawk: *Buteo jamaicensis*

ETYMOLOGY

Buteo: refers to hawk; *jamaicensis*: of Jamaica, origin of first described specimen

Description   In the D.C. area, a typical adult red-tailed hawk reveals these field marks to the observer: a pale chest, a dark-streaked belly band, and the namesake orange-red tail. This is a large, stocky hawk, unlike sleek, long-tailed accipiters such as Cooper's hawk. While soaring, a red-tailed hawk usually shows chocolate-brown bars on the under wings' leading edges. Young birds' tails are finely barred, not colorful.

Size   Because of its short tail, the red-tailed hawk only measures a few inches longer than a crow, though it is a much larger bird. The wingspan is about 49 inches—10 inches beyond that of a crow.

Common Locations   Red-tailed hawks frequent the National Arboretum, the National Mall, Anacostia and East Potomac Parks, the C&O Canal, and other areas. Watch the skies over the city during spring and fall migration periods.

Notes of Interest   The red-tailed hawk is one of the most widespread hawks in North America. Coloration can vary dramatically between individuals and by region. Some pairs nest in the city, but red-tailed hawk numbers get a boost in winter and during the two migration periods, from mid-February to the beginning of April and again from late October through most of November. The somewhat similar red-shouldered hawk is found in the city's larger forested parks year-round, including Rock Creek and Glover-Archbold Parks, at the National Zoo, in C&O Canal National Historical Park, Fort Dupont Park, and at Roosevelt Island. Adult red-shouldered hawks have orange banding underneath and tails boldly banded in black and white.

Ecological Role   A carnivore with a varied diet, the red-tailed hawk hunts rodents, rabbits, and other small mammals, many types of birds, snakes, and sometimes frogs and insects.

KEY POINTS

- Red-tailed hawks often perch out in the open in a tree or on a telephone pole overlooking open habitats at the forest edge. When soaring, they may be seen far from these areas, at times over the city, especially during migration.

Plate 34
RED–TAILED HAWK

- The red–tailed hawk's call is a rasping, almost blood–curdling *screeeeeeee*. But the red–shouldered hawk's *key-ah, key-ah, key-ah* call is more resonant, often heard in the woods or as a pair circles overhead.

Osprey: *Pandion haliaetus*

ETYMOLOGY

Pandion: refers to mythical king of Athens; *haliaetus*: sea eagle; osprey means bone-breaker, a name taken long ago from the bearded vulture or lammergeier of parts of Europe, Asia, and Africa

Description   The osprey stands apart from other raptors. Taxonomists, scientists who classify animals, put this graceful, yet imposing bird in its own family. You may mistake it for a gull as it flaps or wheels overhead on rather slender, curvy wings, but watch an osprey plunge talons first into the water after a fish, or stare one down at its perch in a riverside tree, and there's no mistaking this is a raptor. Ospreys are all white below and brown on back and wings. This combination, along with the white head and dark stripe running around the piercing yellow eye, makes the osprey one of the easiest raptors to identify.

Size   Smaller than a bald eagle and larger than a red-tailed hawk, the osprey averages about 2 feet long with a wingspan of 63 inches.

Common Locations   Ospreys cruise the Potomac and Anacostia Rivers from March to October. Usually one or several pairs nest in city, while others nest just within sight of it. In 2011, a pair placed their bulky stick nest on a construction crane on the shores of the Anacostia. In recent years, they have also nested on the 14th Street railroad bridge passing over the top of Washington Channel (East Potomac Park) and on a pier under the Frederick Douglass Memorial Bridge. Along the Chesapeake Bay, channel markers are a top nest site choice. A pair usually nests on a marker in the Potomac near Daingerfield Island, just south of Ronald Reagan Washington National Airport, within sight of the city. Ospreys winter from the Carolinas south to South America.

Notes of Interest   The Chesapeake Bay watershed, which includes Washington, D.C., hosts one of the largest osprey populations in the world.

Ecological Role   A fish eater, the osprey dives feet-first for a variety of species. It can handle fish up to a foot long.

KEY POINTS

- For such an impressive bird, the osprey vocalizes in surprisingly high-pitched chirps and whistles.
- Bald eagles sometimes rob ospreys of their catches.
- The osprey is found on all continents except Antarctica.

Plate 35

OSPREY

American Coot: *Fulica americana*

ETYMOLOGY

Fulica: refers to waterfowl; *americana* differentiates it from other coots elsewhere in world

Description The American coot resembles a duck in size and general shape. But it's a close relative of rails and gallinules. The coot's bill is cone shaped, not flattened, and is whitish with a narrow black band. Unlike most ducks, the coot's body is one color, soot gray. Much of a coot's body floats above the water; ducks sit lower. Also, a coot moving along in the water bounces its neck in a slight "chugging" motion. When a coot takes to the air it seems to be working hard; the bird's flight is weak and its feet skip across the water's surface for a little while before the bird is aloft.

Size At about 15 inches long, the American coot is smaller than a mallard but larger than a green-winged teal, the smallest dabbling duck.

Common Locations Coots visit the Potomac River, usually in flocks of six to a dozen or more on open water near the shoreline. The Tidal Basin and Hains Point often have some. They are also seen at Georgetown Reservoir and Constitution Gardens, and along the George Washington Memorial Parkway. Over the past decade, birders have reported a drop in coot numbers in the city.

Notes of Interest Scan coot flocks to find interesting wild ducks that often accompany them.

Ecological Role Coots are scrappy birds that find various foods in a variety of ways. They skim algae from the water's surface, walk along the shore nibbling stems, leaves, and seeds, tip up their rear ends to seek food in the shallows, or dive down in deeper waters. In addition to plant matter, coots eat fish, worms, tadpoles, crustaceans, and other small animals. They may even steal food from nearby ducks.

KEY POINTS

- In Washington, D.C., coot numbers peak in November and December. Some flocks remain through the winter and then numbers peak again in March before dwindling by late April.
- Unlike ducks' webbed feet, coots have wide, rounded (lobed) feet that help them paddle in water and walk on land.
- American coots utter a variety of sounds, including grunts, clucks, and trumpet-like noises.

Plate 36

AMERICAN COOT

Ring-billed Gull: *Larus delawarensis*

ETYMOLOGY

Larus: gull; *delawarensis* refers to the Delaware River, where first described

Description   The city's most numerous gull is often very easy to observe without binoculars. Adults have bright yellow legs and a yellow bill that appears wrapped with a single jet-black band. The primary feathers are black tipped with a few white spots, while the back, or mantle, is light gray. Otherwise, adult birds have white heads and underparts. Young birds in their first winter have gray backs but are more blackish on the wings and are blotched in brown. Unlike adults, they have pinkish legs and pinkish bills tipped with black.

Size   Smaller than the herring and great black-backed gulls with which it associates, the ring-billed gull measures about 18 inches long, with a wingspan of 4 feet. It is much larger than the Bonaparte's gull that can be seen coursing up the Potomac from late March to mid-April. (That bird has distinctive white wing patches and in breeding plumage a black head.)

Common Locations   In winter, Hains Point and some other spots along the Anacostia and Potomac Rivers host large numbers of ring-billed and other gulls. Ring-billed gull flocks loiter on and wheel over the National Mall much of the year, except late spring and early summer, when most migrate north to nest. A few wheeling ringbills often appear in news footage showing a U.S. Capitol or National Mall backdrop.

Notes of Interest   Although a common bird in Washington, the ring-billed gull nests around the Great Lakes, across southern Canada, and in the northern parts of the prairie states.

Ecological Role   Ring-billed gulls do not restrict themselves to watery habitats. Consummate opportunists and omnivores, they find food almost anywhere they roam, from trash at dumps, to dead fish on the shore, to grubs in plowed fields, and insects and earthworms in the grass. They will also eat grains.

KEY POINTS

- The best way to start tackling one of birding's greatest challenges—gull identification—is to get to know all the field marks of the most common, the ring-billed gull.
- Adult herring gulls have similar mantle colors, but they have pink legs and are much larger. They also have a pink spot, instead of a ring, on their longer yellow bills.

Plate 37
RING–BILLED GULL

- Gull numbers spike in winter during hard freezes, when gulls concentrate at remaining areas of open water. Hains Point is a particularly good place to sort through large numbers of common gulls—ring-billed, herring, and great black-backed gulls—in hopes of finding a lesser black-backed or rarer Iceland or glaucous gull.

Mourning Dove: *Zenaida macroura*

ETYMOLOGY

Zenaida: honors the first name of French ornithologist Charles-Lucien Bonaparte's wife; *macroura*: long-tailed

Description   On close inspection, this is a beautiful bird: fawn-brown coloration with a smattering of large black spots on the back and wings, a wash of pink with metallic green scaling on the neck, a frosting of gray on crown and neck, and bright pink legs and feet. In flight and at rest, the mourning dove's long, tapered, white-edged tail is a reliable field mark.

Size   A foot long, with a wingspan of a foot and a half.

Common Locations   Within the city, this bird frequently forages on the ground, perches on wires, and nests and roosts in a wide variety of trees and shrubs. The National Mall and the National Zoo are among the many places where these birds nest.

Notes of Interest   The mourning dove is the most widespread and abundant native dove in North America. (Rock or street pigeons were introduced.)

Ecological Role   The mourning dove is a seedeater, searching the ground for grass seeds, weed seeds, grains (such as cracked corn), and sunflower and other seeds at feeders.

KEY POINTS

- The "sad" cooing of this dove gives the bird its name. This sound is a familiar part of the morning bird chorus throughout the city from spring through summer, as birds court and set up territories.

- The larger rock, or street, pigeon, the only other dove regularly seen in the city, comes in varied color phases, from white to red orange to blue and gray. Although still rare in the region, the introduced Eurasian collared-dove will likely extend its range north and east in coming years, becoming the city's third breeding dove.

- In Washington, D.C., mourning doves may nest up to three times per year, sometimes starting as early as February and ending as late as October. Nesting season peaks in late March and April. The males' hawk-like soaring courtship flights cause double takes among birders. Mourning doves build flimsy, flat-profile stick nests out of which the ungainly young seem likely to tumble, although they usually manage to stay put.

Plate 38

From top: CHIMNEY SWIFT & MOURNING DOVE

Chimney Swift: *Chaetura pelagica*

ETYMOLOGY
Chaetura: bristle tail; *pelagica*: marine

Description   In his pioneering field guides, Roger Tory Peterson referred to this ashy gray bird as "a cigar with wings." Many visitors to Washington between April and October mistake the dark chimney swift for a day-flying bat, as it teeters and wheels over the city's buildings and streets. The only member of its tribe found in the East, the chimney swift's all-dark coloration, scimitar-shaped wings, short, nonforked tail, and faster, more frenetic flight separate it from swallows, which inhabit the city during the same time of year. Learn the distinctive chittering of a chimney swift flock and you will frequently find the birds circling high above the city, where they snap up flying insects.

Size   At just over 5 inches long, the chimney swift is a bit longer than a goldfinch but with a wider wingspan of 14 inches.

Common Locations   Chimney swifts, true to their name, often roost and nest in chimneys, although their traditional nesting sites are in hollow trees. Mid-April to mid-October, they can be seen anywhere over the city.

Notes of Interest   Expect to see chimney swifts only in flight. With short legs and hook-like toes, they only cling to vertical surfaces, such as the brick inside chimneys. Swallows, however, often perch on wires and on dead tree limbs.

Ecological Role   Chimney swifts eat flying insects, including flies, beetles, bugs, and flying ants.

KEY POINTS

- During migration, large numbers of chimney swifts may roost together in a choice chimney, where hundreds drop in at dusk and fly out at dawn.
- Chimney swifts craft their wall-clinging nests out of bits of twig and their own sticky saliva.
- In the late 1800s, ornithologists believed this bird wintered in Mexico or in Central America. South America is now the known destination.

Chimney Swift

Downy Woodpecker: *Picoides pubescens*

ETYMOLOGY

Picoides: resembling a woodpecker; *pubescens*: refers to downy hairs

Description   This petite woodpecker is the city's and the country's smallest. Its coloration recalls that of a penguin, mostly black above and all white below. The male sports a scarlet blotch on his nape. When a downy woodpecker clings to a tree trunk with its back to you, you can see a large white oval on its black back. Two broad white stripes adorn the face and white bands and squares decorate the otherwise black wings. (See below for comparison to similar hairy woodpecker.)

Size   At 6.75 inches long, the downy woodpecker is slightly longer than a tufted titmouse or a house sparrow.

Common Locations   The downy woodpecker is a common year-round fixture in the city's wooded neighborhoods, parks, and gardens.

Notes of Interest   The hairy woodpecker is the downy's larger cousin. Found in forests (Rock Creek Park and along the River Trail at Kenilworth Aquatic Gardens, for example), it also sometimes visits backyards with large trees. Both species share almost identical markings but size, proportions, and voice differ. The downy utters a thin *pihk* call. The hairy's call is a sharp *PEAK!* The downy also has an accelerating, descending rattle, while the hairy's rattle is more harsh and on one pitch. The downy woodpecker's bill is stubby and almost nail-like, about half the head length. The hairy's more formidable bill appears almost as long as its head.

Ecological Role   With its small body and the smallest woodpecker bill around, the downy usually doesn't dig very deep for food. It will venture where many other woodpeckers dare not go—on very thin young trees and weed stalks. Downys are often seen at bird feeders, although their primary food is insects. They also eat small fruits such as mulberries.

KEY POINTS

- With an abundance of trees, Washington is a capital city for woodpeckers. Downy and red-bellied woodpeckers and northern flickers are usually common. With some looking in the parks, pileated, hairy, and (fall through spring) yellow-bellied sapsuckers often turn up. Only the red-headed woodpecker requires a good amount of luck.

Plate 39

DOWNY WOODPECKER

- Downy woodpeckers often join mixed feeding flocks that may include chickadees, titmice, warblers, kinglets, creepers, and nuthatches.

Red-bellied Woodpecker: *Melanerpes carolinus*

ETYMOLOGY

Melanerpes: refers to black creeper; *carolinus*: of the Carolinas

Description   A zebra-striped back and a generous splash of scarlet on the hindneck set this bird apart from other local woodpeckers. On the male, bright red runs from nape to bill. In the female, the red is only on the back of the neck and on a small reddish-yellow patch just above the bill. Underparts are whitish.

Size   At just over 9 inches long, the red-bellied woodpecker is a bit longer than a cardinal.

Common Locations   Red-bellied woodpeckers inhabit forests and Washington, D.C., neighborhoods with large trees. They often visit feeders. Rock Creek Park, Glover-Archbold Park, the C&O Canal, the National Arboretum, and other forest parks ring with their calls.

Notes of Interest   In the late 1800s, ornithologists wrote that the red-bellied woodpecker nested in eastern states from Florida to Maryland. Since then, this adaptable woodpecker has extended its range north to southern Ontario and southern New Hampshire.

Ecological Role   A dazzling midsized woodpecker usually seen hitching up trees, the red-bellied sometimes drops to the ground to snatch seeds, acorns, insects, or berries.

KEY POINTS

- The striking *cleeurr* of the red-bellied woodpecker is one of the city's most distinctive bird sounds.
- Watch very closely and you may glimpse the namesake red feathers on the bird's belly.
- The more apt name for this bird is already assigned to another: The red-headed woodpecker, a locally scarce species with an entirely red head, occasionally visits Rock Creek Park and the other forested parks during migration.

Plate 40

RED-BELLIED WOODPECKER

Northern Flicker: *Colaptes auratus*

ETYMOLOGY

Colaptes: refers to pecking or chiseling; *auratus*: gilded, referring to this bird's yellow wing linings

Description   The northern flicker is a feast for the eyes, an unusual mix of spots, lines, and color. Large black spots stipple the beige underparts, bordered above by a black chest crescent. Unlike the red-bellied woodpecker's zebra-striped back, the flicker's is brown striped with black. The nape is gray inset with a scarlet wedge, and the bill is long and black. In flight, the northern flicker is easy to identify: Yellow wing flashes and a white rump are bold field marks. Want to know a flicker's gender? Males have a black "moustache" behind the bill that females lack.

Size   The flicker is larger than a red-bellied woodpecker by about 3 inches and smaller than a pileated woodpecker by 4 inches.

Common Locations   Rock Creek Park, the National Zoo, Hains Point, Anacostia Park, and the National Arboretum are among the places to look for this attractive bird, which also nests in some of the city's well-treed neighborhoods.

Notes of Interest   The "yellow-shafted" form of this species—recognized in part by the yellow undersides of its wings—lives in the eastern United States. The "red-shafted" form, with its salmon wing linings, is found in the West. These were once considered separate species. The two often hybridize where their ranges meet on the Great Plains and in other central areas.

Ecological Role   Ants constitute a large part of the diet, but flickers also eat many other insects, including beetles and termites. Flickers also eat some berries, including black cherry fruits.

KEY POINTS

- In spring and early summer, the flicker utters a long, steady series of choppy but uniform notes: *dutdutdutdutdutdutdutdutdutdut.* The flicker also makes a distinctive "teeyr" call.
- The northern flicker is not just a woodland bird. It often feeds in open grassy areas near scattered large trees, particularly in warm months, when insects are most active.
- Flickers can be found in Washington year round, but spring and fall bring large numbers of migrating birds.

Plate 41

NORTHERN FLICKER

Pileated Woodpecker: *Dryocopus pileatus*

ETYMOLOGY

Dryocopus: refers to a tree-dagger; *pileatus*: capped

Description   The city's largest woodpecker is not likely to be misidentified. Unlike other Washington woodpeckers, the pileated has a pointed scarlet crest. When seen clinging to a tree, this crow-sized bird appears mostly black, with thick stripes of white and black on the face and neck. In flight, its strong, stroking wing beats and bold black-and-white wing pattern are distinctive. The male pileated woodpecker has a red moustache and forehead; a female has a blackish forehead and black moustache.

Size   The pileated woodpecker is 16 to 18 inches long with a wingspan of about 29 inches—far larger than the city's other woodpeckers.

Common Locations   The pileated woodpecker nests in mature woodlands. The C&O Canal National Historical Park, Rock Creek Park, Glover-Archbold Park, Roosevelt Island, Kenilworth Aquatic Gardens' riverside woods, and the National Zoo are among the places to find this bird.

Notes of Interest   Despite a similar appearance, the legendary, probably extinct, ivory-billed woodpecker is not a close relative of the pileated woodpecker. Close relatives include the lineated woodpecker, found from Mexico to northern Argentina, and Europe's black woodpecker.

Ecological Role   Pileated woodpeckers hack deep into rotting wood, leaving telltale vertical rectangles resembling trail blazes. They hack off bark strips in search of grubs (beetle larvae) and may be seen on the forest floor or edge, working stumps and fallen trees for one of their favorite foods, carpenter ants, as well as termites and other insects. They will also clamber onto thinner branches to reach berries such as black cherry or wild grape. Sometimes they eat beechnuts and acorns.

KEY POINTS

- The pileated woodpecker is often heard before it is seen thanks to the bird's loud, irregular "kuk-kuk" calls and a jarring series of piping notes.
- Pileated woodpeckers often leave the deep woods to investigate tall trees in well-shaded Washington neighborhoods.

Plate 42

PILEATED WOODPECKER

Eastern Kingbird: *Tyrannus tyrannus*

ETYMOLOGY

Tyrannus: refers to a tyrant or ruler

Description   Kingbirds provide a flying study in contrasts: They are black above and white below, with a black tail broadly banded in white at the tip. In flight, eastern kingbirds fly with stiff, short wing beats. They usually fan their tails when displaying or landing. Although only present for about four months, chattering, pugnacious kingbirds seem to have always been a part of the scenery once they return in spring.

Size   At 8.5 inches, the eastern kingbird is a bit smaller than a cardinal.

Common Locations   The first eastern kingbirds show up in late April. Most leave the city by early September. They are the quintessential park bird, nesting in scattered tall shade trees by streams and wide open grassy areas such as playing fields and picnic grounds. Look for them in Rock Creek Park, at Anacostia Park, along the C&O Canal, at Kenilworth Aquatic Gardens, and many other places.

Notes of Interest   Eastern kingbirds are a fixture throughout the eastern United States and Canada but also range as far west as eastern Oregon, Washington, and British Columbia, even reaching the southern tip of the Yukon Territory. Territorial, insect-eating birds in summer, they range across Amazonia in winter in silent, fruit-eating flocks.

Ecological Role   During spring, summer, and early fall, eastern kingbirds mostly eat insects, including wasps, bees, flies, flying ants, grasshoppers, and beetles. They also consume berries, including black cherry. Their winter diet in South America includes small rain forest fruits.

KEY POINTS

- Unlike many other flycatcher family members, the eastern kingbird lives out in the open, perching on treetops, backstops, signs, and wires.
- Named for its aggressive territorial habits during spring and summer, the kingbird will even attack hawks near its nest.
- Other flycatchers common in the city's parks are harder to see: the great crested flycatcher, eastern wood-pewee, and Acadian flycatcher are easy to hear but harder to glimpse. The eastern phoebe, a grayish, tail-pumping bird, is often easy to see near bridges over streams, a favored nesting location.

Plate 43

EASTERN KINGBIRD

Blue Jay: *Cyanocitta cristata*

ETYMOLOGY

Cyanocitta: dark blue chattering bird; *cristata*: crested

Description   The blue jay is one of the city's most colorful birds. Males and females are colored alike: rich light blue on crown, crest, and back, with more vivid bright blue wings checkered in black and white. The tail is also blue, but with black bars. In flight, blue jays flash white on the trailing edges of their wings and their tail corners.

Size   At 11 inches long, blue jays are an inch shorter than mourning doves. These two birds, along with rock or street pigeons, are usually the largest birds at feeders in the city.

Common Locations   Watch for them in any park or neighborhood with large trees. Oaks produce acorns, a favored food. Bird feeders bring these flashy birds low and out in the open.

Notes of Interest   Jays belong to the crow family.

Ecological Role   Branded as notorious nest raiders, blue jays do sometimes eat other bird's nestlings and eggs, but these form a small part of the diet. When available, acorns, beechnuts, chestnuts, and hickory nuts are important food items. From an ecological perspective, blue jays are like winged Johnny Appleseeds, often flying long distances to cache acorns or other tree nuts, some of which germinate when the jay neglects to retrieve them. Blue jays also eat peanuts, sunflower seeds, suet, and cracked corn put out at feeders, and wild fruits from vines, shrubs, and trees such as black cherry. Insects and other invertebrates make up at least a fifth of the diet.

KEY POINTS

- Blue jays are found year round in the city, but the bird you see in winter might not be the same one that nests in the same spot in summer. Late April to May and late September into October bring an influx of migrating birds. During this time, migrating flocks can be seen flapping over the treetops.

- When blue jays store, or cache, acorns and other nuts, they often hide them by covering them with leaves or pebbles. Each fall a blue jay may cache several thousand acorns.

- The blue jay makes a number of calls, but the piercing *jay* is the most familiar. Blue jays often imitate red-shouldered hawks, so some care should be taken when hearing the loud *key-ah, key-ah, key-ah* of an apparent hawk. It may just be a jay fooling you.

Plate 44

BLUE JAY

Carolina Chickadee: *Poecile carolinensis*

ETYMOLOGY

Poecile: referring to pied, or black and white, coloration; *carolinensis*: of the Carolinas

Description   Chickadees wander through the trees in small groups, constantly calling and clinging almost upside down as they examine tree branches for seeds, berries, and tiny invertebrates. Both sexes sport black cap and bib and have white cheeks, both in contrast to the gray-brown backs and wings. The underparts are faded yellowish beige.

Size   At a shade less than 5 inches long, the Carolina chickadee is about as long as a typical warbler and 1.5 inches shorter than a house sparrow.

Common Locations   Well-wooded parks and neighborhoods host chickadee flocks. Carolina chickadees frequent feeders, such as those at the Rock Creek Park nature center.

Notes of Interest   Carolina chickadee pairs tend to stay together, splitting off from flocks in early spring. After nesting, pairs become flock members again by fall.

Ecological Role   Found year round in the city, the tiny but mighty chickadee toughs out winter by eating seeds and berries and scouring woody surfaces to find dormant invertebrates. In warmer months, it eats a variety of insect life, including caterpillars.

KEY POINTS

- To find other small birds, including titmice, creepers, nuthatches, kinglets, wrens, warblers, and downy woodpeckers, watch and listen for chickadee flocks. They often travel in mixed company.

Plate 45

CAROLINA CHICKADEE

Tufted Titmouse: *Baeolophus bicolor*

ETYMOLOGY

Baeolophus: having a small crest; *bicolor*: two colored

Description   This pencil-gray bird is easy to identify thanks to its small cardinal-like crest, beady black eyes, and orange flanks.

Size   At 6.5 inches long, it is about as long as a house sparrow but a bit longer tailed and not as stocky.

Common Locations   Often found in the company of slightly smaller chickadees, tufted titmice frequent wooded areas, including Rock Creek Park, the C&O Canal, Glover-Archbold Park, the National Zoo, and the National Arboretum. They also frequent bird feeders.

Notes of Interest   Like chickadees, titmice live year round in Washington, D.C., often flocking with a changing cast of characters. During much of the year, flock mates include wrens, downy woodpeckers, and white-breasted nuthatches. In fall and winter, kinglets, creepers, and yellow-rumped warblers join the mix. In spring and early fall, a variety of migrating warblers add a colorful accent to the groups.

Ecological Role   Tufted titmice eat many different types of insects, as well as spiders and snails. Invertebrates make up about 60 percent of their diet. In winter, they eat dormant invertebrates when they can find them but rely more on seeds, nuts, and fruits.

KEY POINTS

- The tufted titmouse's song is among the easiest to learn: a clear *peter-peter-peter-peter*.
- At feeders, titmice fly off with single sunflower seeds, which they hold between their feet and hammer open with their small black bills.
- Tufted titmice and Carolina chickadees nest in small cavities in trees. Sometimes, they nest in boxes put out to attract nesting bluebirds.

Plate 46

TUFTED TITMOUSE

White-breasted Nuthatch: *Sitta carolinensis*

ETYMOLOGY

Sitta: a nuthatch; *carolinensis*: of the Carolinas

Description   A steely blue-backed bird that clambers head first down tree trunks, calling a nasal *yank yank yank*. Males have jet-black caps and white faces; females have dark gray caps and white faces. Below, white-breasted nuthatches, true to their name, are white, but under their tails (at the vent) the feathering is rust colored.

Size   At 5.75 inches long, the white-breasted nuthatch is a little longer than a goldfinch.

Common Locations   Parks and well-treed suburbs harbor white-breasted nuthatches. They usually nest in cavities in mature trees and are found in small numbers in Rock Creek Park, Glover-Archbold Park, the National Arboretum, and many other sites.

Notes of Interest   The white-breasted nuthatch gets its common name from its habit of placing sunflower seeds and soft nuts in a cranny and hacking them open with its chisel-like bill.

Ecological Role   White-breasted nuthatches feed on tree nuts such as beechnuts and acorns, sunflower seeds, suet, and grains, which constitute much of their fall and winter diet. In warmer months, they capture beetles, caterpillars, ants, and other invertebrates.

KEY POINTS

- This species is only likely to be confused with the red-breasted nuthatch, a smaller species sometimes present fall to early spring at the National Arboretum, Lyndon B. Johnson Memorial Grove, and other parks rich in conifers. The red-breasted nuthatch is orangish below, has a thin black line through the eyes, and usually forages along branches, not on trunks.

- Unlike the red-breasted nuthatch, which appear unpredictably from the north, white-breasted nuthatches do not wander far and can be found year round.

- Bird feeders likely boost white-breasted nuthatch populations, and those of tufted titmice, red-bellied and downy woodpeckers, and cardinals. At the very least, bird feeding, along with the impressive amount of forest and park within the city, helps keep populations stable.

Plate 47

WHITE-BREASTED NUTHATCH

Carolina Wren: *Thryothorus ludovicianus*

ETYMOLOGY

Thryothorus: refers to rushing through reeds; *ludovicianus*: refers to French king Louis XIV, for whom the Louisiana Territory was named

Description Noisy and always on the move low in brush or vine tangles, the Carolina wren is one of Washington's most familiar woodland birds. A somewhat bulky-headed bird with a distinctive white eye line, it usually keeps its tail cocked at an angle, like many other wrens do. Rich rust adorns the back and wings, while the underparts are yellow orange. The Carolina wren's bill differs from that of sparrows, chickadees, titmice, and warblers—it is long, thin, and slightly down-curved. Boisterous and active, Carolina wrens are easy to hear, but it often takes some patience to get a good look at them.

Size At 5.5 inches long, the Carolina wren is a bit longer than a Carolina chickadee, yet shorter than a tufted titmouse.

Common Locations Carolina wrens nest in brushy backyards, but they are most abundant in dense tangles that grow at the forest edge or within woodlands. Roosevelt Island, Rock Creek Park, Glover-Archbold Park, Dumbarton Oaks, the National Arboretum, and the C&O Canal are among the many places they are found.

Notes of Interest Two smaller wrens frequent Washington, D.C., at different seasons. The mouse-brown house wren nests in the city from April into the summer but is usually gone by October. The equally mouse-like winter wren appears October to April, favoring fallen logs, brush piles, and other shelter. This bird has dark flanks and a shorter, stubby tail.

Ecological Role Carolina wrens eat many invertebrates, including a wide variety of insects, spiders, and on occasion snails. They sometimes eat seed at feeders in winter, when berries and suet provide other important food sources.

KEY POINTS

- Carolina wren pairs stay together all year, and small family groups are often seen. In very snowy or icy winters, the local Carolina wren population may plummet, only to rebound within a few years.

Plate 48

CAROLINA WREN

- The Carolina wren's song, *teakettle teakettle teakettle*, or *turtle-lee, turtle-lee, turtle-lee*, can be heard in the woods much of the year. The birds also make a few grating calls.
- Carolina wrens sometimes choose quirky nesting sites. They may stuff their twiggy, grassy nests into old shoes, potted plants, or cracks in buildings.

American Robin: *Turdus migratorius*

ETYMOLOGY

Turdus: a thrush; *migratorius*: wandering

Description   The American robin is one of North America's most familiar songbirds thanks to its ability to thrive in human-made habitats. Both sexes have bright orange underparts and grayish upperparts, but the male has a blackish head. The female's head color matches that of her back. The feathering under the tail (the vent) is white, as is a wide broken ring around the eye. The thick, long bill is yellow. Juvenile birds are heavily blotched in black below, with more subdued whitish streaks and spots above.

Size   The American robin is a medium-sized songbird, 10 inches long with a wingspan of 17 inches.

Common Locations   American robins are found year round in Washington, D.C., and nest in open spaces with scattered trees, including neighborhoods, parks, and the National Mall.

Notes of Interest   Although both American and European robins belong to the thrush family, the American robin is more closely related to Europe's blackbird (*Turdus merula*).

Ecological Role   The American robin changes its diet to match food sources available at different seasons. Late fall through early spring, these birds flock to berry-bearing plants. In spring and summer, they frequent lawns, snatching earthworms and other invertebrates, while still watching for ripe mulberries and other small fruits.

KEY POINTS

- In spring and early summer, the city's dawn chorus is dominated by American robins singing their territorial song, *chirrup cheree, chirrup cheree, chirrup cheree.*
- The robin's nest—a loose cup of grasses, twigs, and other materials plastered together with dried mud—is one of the most-often seen nests in the area. Robins often place their nests not far above head level on a horizontal branch.

Plate 49

AMERICAN ROBIN

Gray Catbird: *Dumetella carolinensis*

ETYMOLOGY

Dumetella: refers to little thicket; *carolinensis*: of the Carolinas

Description   Both male and female are dark gray, perfect for hiding in the shadows. The have black caps and black eyes. The vent, or area under the tail, is chestnut colored.

Size   The catbird measures about 8.5 inches long, about as long as a cardinal.

Common Locations   Shady backyards and probably all of the city's forested parks host gray catbirds from April through October. Catbirds winter from the East Coast down to Central America. Sometimes, one or two ride out the winter in the city. Gray catbirds at the National Zoo and on the National Mall are tame. They hop onto benches, drop into trashcans for morsels or insects, and feed their young in the gardens.

Notes of Interest   Well-described both in scientific and common names, the gray catbird is, as its genus name states, common in tangles of shrubs, small trees, and vines. There, its cat-like calls draw the attention of those who care to listen.

Ecological Role   In spring and summer, insects and spiders make up much of the diet. In summer and fall, gray catbirds eat berries they pluck from vines, shrubs, and trees. Gray catbirds, as well as mockingbirds, robins, starlings, and others, disperse the seeds of these plants via their droppings.

KEY POINTS

· Catbirds frequently nest in shrubs and small trees next to houses, unbeknownst to the human residents there.

· In addition to its namesake *meew* call, the catbird makes a machine-gun-like *tat tat tat tat tat* and a low *quat*. The song is a squeaky mockingbird-like performance, often punctuated with mews.

Plate 50

GRAY CATBIRD

Northern Mockingbird: *Mimus polyglottos*

ETYMOLOGY

Mimus: a mimic; *polyglottos*: many-tongued

Description   The northern mockingbird is a medium-sized songbird with a flare for attracting attention. Although colored a somber gray, darker above and whitish-gray below, in flight the wings and tail flash white, thanks to the bird's bold wing bars and outer tail feathers. At rest, the tail appears long, narrow, and black, except for white outer tail feathers. The eye is pale yellow.

Size   From its thin bill to its slender tail, the northern mockingbird measures 10 inches long, about as long as an American robin, although the mockingbird is longer of tail and shorter of body.

Common Locations   Open parkland, roadsides, and backyards are the domain of northern mockingbirds, which can be found all year long in Washington. From March to midsummer, they are very vocal, and sometimes sing at night (particularly unmatched males). By August, they may be harder to find during the quiet period between the fledging of their young and the establishment of fall and winter feeding territories. Watch for them on the National Mall, around the monuments, and singing from streetside trees.

Notes of Interest   The brown thrasher, the mockingbird's slightly larger country cousin, passes through the city from about mid-March through April and again from late August through October. Some nest in the city in brushy parts of Rock Creek Park, Kenilworth Aquatic Gardens, the National Arboretum, Fort Dupont, and perhaps along the C&O Canal and in Anacostia Park. The thrasher is long tailed and rich reddish rust above and white below with black streaking. It has a long, slightly down-curved bill.

Ecological Role   The mockingbird uses its long, slender bill to grub in the soil and grass for invertebrates, from beetles to sowbugs to snails, and to pluck berries from trees, bushes, and vines. In winter, mockingbirds eat raisins, peanut butter, and suet at feeding stations. Their diet is more or less half plant and half animal matter, with the emphasis on the latter during warm months when such foods are both abundant and critical to the growth of their nestlings.

KEY POINTS

- The mockingbird's namesake habit of imitating other birds and other sounds can make for interesting listening. Repertoires change and grow each year. The city's mockingbirds frequently

Plate 51

NORTHERN MOCKINGBIRD

imitate Carolina wrens, blue jays, woodpeckers, and many other species, sometimes along with everyday sounds such as car alarms.

- When defending nesting territories, northern mockingbirds sing from wires, treetops, and buildings. They chase away other birds and sometimes dive-bomb cats and pedestrians that pass too close to their hidden nests.
- Mockingbirds' showy behaviors often draw a viewer's attention. The birds run across lawns like little roadrunners then stand still and flex and flash their white-splashed wings. For decades, ornithologists have sparred over why these birds flash their wings: A feeding strategy? A sign of wariness? Territorial display? You may come up with your own hypothesis.

Yellow Warbler: *Dendroica petechia*

ETYMOLOGY

Dendroica: tree dwelling; *petechia*: refers to male's reddish breast streaking

Description   The yellow warbler is aptly named. Wings, back, head, underparts, and tail are all yellow. The male has orange-red streaks across his breast and down his flanks. Many warblers have white tail spots, but the yellow warbler's are, aptly, yellow.

Size   The yellow warbler measures 5 inches long, with a wing-span of 8 inches.

Common Locations   Yellow warblers pass through most wooded parks and many backyards during migration. But the yellow warbler also nests in a few places, particularly those with low, watery areas dominated by thickets of black willow and other small trees and shrubs. The boardwalk trail at Kenilworth Aquatic Gardens is a good place to see them close up, as well as East Potomac Park.

Notes of Interest   During spring and fall migration, there are more than 30 warbler species that may be seen in the city. Some are more common than others are. Along with the yellow and yellow-rumped warblers and common yellowthroat described here, these eight species are most often seen: northern parula (which nests), chestnut-sided, magnolia, black-throated blue, black-throated green, blackpoll, and black-and-white warblers, and American redstart.

Ecological Role   The yellow warbler feeds mainly on insects, with a particular emphasis on various types of caterpillars found in the trees.

KEY POINTS

- The yellow warbler is the most widespread warbler in North America, with more than 40 described subspecies.
- In spring, its clear and loud song is among the most distinctive: It seems to say *Sweet sweet sweet sweet, I'm so sweet.*

Yellow-rumped Warbler: *Dendroica coronata*

ETYMOLOGY

Dendroica: refers to tree dwelling; *coronata*: crowned

Description   The yellow-rumped warbler is a bit sluggish compared with other warblers, and thus easier to view. During peak migration, it is usually the most common species as well, and it's the most likely to be seen in winter. At all seasons, the bright yellow

Plate 52
Clockwise from top: YELLOW WARBLER,
YELLOW-RUMPED WARBLER, COMMON YELLOWTHROAT

rump stands out, and you can usually see a yellow wash on the sides, below the wings. In spring migration, the male is as dapper as any other warbler and the somewhat less colorful female is not far behind. In breeding plumage, the yellow crown can be seen and the yellow sides in both male and female stand out more because they are outlined in black, while the white throat is fringed by a black mask above and black-streaked chest below. Spring birds also have black-streaked, slate gray backs in place of the drab brown non-breeding plumage.

Size Yellow-rumped warblers measure about 5.5 inches long, with a 9-inch wingspan.

Common Locations Yellow-rumped warblers often show up in mixed songbird flocks in parks such as Rock Creek Park, Kenilworth Aquatic Gardens, and the C&O Canal National Historical Park. The National Arboretum and LBJ Memorial Grove are good places to watch for them in winter.

Notes of Interest From November through much of March, the yellow-rumped warbler is the only warbler regularly seen in Washington, D.C., though even they are scarce during those months. The pine warbler may turn up at the National Arboretum (where it nests in spring) and at LBJ Memorial Grove. Palm, orange-crowned, or Nashville warbler, or common yellowthroat, put in rare winter appearances.

Ecological Role Creeping along branches, hovering to snatch flying insects, and climbing through vines, the yellow-rumped warbler is adapted to feeding in a variety of ways on a variety of foods. In warmer months, most of the diet is insects or spiders but during fall and winter, these birds gulp down small berries of poison ivy and other plants.

KEY POINTS

- Yellow-rumped warblers usually call *chek* while flying or foraging. In spring, their song is often heard, a sweet but not punchy *swiddle swiddle swiddle swiddle swiddle sweetl sweetl sweetl sweetl*.
- Most other warblers winter in Mexico, Central America, the Caribbean, or northern South America.

Common Yellowthroat: *Geothlypis trichas*

ETYMOLOGY

Geothlypis: earth finch; *trichas*: refers to song thrush.

Description   Seeing a male sing in the open is a treat: His throat and chest are an intense lemon yellow, and his eye is concealed by a broad black mask, fringed above in white. The female is much more somber in color but like the male has a yellow throat. Both sexes are tan olive on their wings and back and keep low in the vegetation, often going undetected unless the observer seeks them out.

Size   At just 5 inches long, with an almost 7-inch wingspan, the common yellowthroat is about the size of a chickadee.

Common Locations   Unlike many warblers, which generally stick to the treetops, the common yellowthroat sings, feeds, and nests in rank, low vegetation in unmowed fields, at the forest edge, and in wetlands. They are not found in manicured gardens or backyards, except during migration. Among the places to look for them: Kenilworth Aquatic Gardens, Roosevelt Island, and the C&O Canal.

Notes of Interest   The common yellowthroat acts more like a wren than a typical warbler. It cocks its tail like a wren, and it typically feeds in dense clumps of vegetation near the ground. The somber-hued female and young are mostly olive colored and can be mistaken for wrens, especially in shadow.

Ecological Role   The insectivorous common yellowthroat retreats in winter to southern states, Mexico, Central America, and the Caribbean. While on its breeding grounds, it eats a wide variety of small invertebrates, including damselflies, caterpillars, grubs, and grasshoppers.

KEY POINTS

- Listen for the yellowthroat's loud *wichety-wichety-wichety* song. Its loud *chip* note is also distinctive from many other warblers' chips but takes time to learn.
- The common yellowthroat is one of the most responsive species when an observer makes "spishing" sounds. Males and females will often pop into view on a stalk or twig to try to see the noisemaker. However, care should be taken not to make such noises if it is nesting season, as this may stress the birds.
- Common yellowthroats are found in Washington, D.C., from mid-April to mid-October, although a few individuals may linger well into winter.

Red-winged Blackbird: *Agelaius phoeniceus*

ETYMOLOGY

Agelaius: of a flock; *phoeniceus*: refers to dull scarlet color

Description   One of North America's most abundant birds, the red-winged blackbird is typically seen in marshy habitat. From early spring into summer, territorial males sing from atop cattails and other obvious perches, lowering their wings, fluffing out their scarlet shoulder patches, and belting out a rusty-hinge-like *Konk-a-Ree!* The male is easy to identify: all black but for its red shoulder bordered below in pale yellow. The female looks very different: Streaked all over in brown, she resembles a large, pointy-billed sparrow.

Size   Just below 9 inches long, the red-winged blackbird is about the size of a cardinal.

Common Locations   These birds show up at backyard feeders and in fall and winter, large numbers join starlings, grackles, and cowbirds visiting cornfields and other cultivation. During breeding season, red-winged blackbirds set up shop in wet areas large and small. These may be marshes, such as those at Kenilworth Aquatic Gardens and Roosevelt Island, or small cattail-choked drainage pools dug at construction sites.

Notes of Interest   The vigilance of male red-winged blackbirds on territory may protect yellow warblers that nest nearby from marauding brown-headed cowbirds.

Ecological Role   Much of the red-winged blackbird's diet consists of grass and weed seeds and grain left in fields or at feeders. In warmer months, however, at least half the diet consists of insects and other invertebrates.

KEY POINTS

- Red-winged blackbirds are polygynous, meaning multiple females will nest within a single male's territory. But this doesn't mean that the females only mate with the male defending that territory.
- During peak nesting season, a male may spend 75 percent of his time defending territory—singing and chasing off intruders.
- Male red-winged blackbirds may briefly ride the backs of herons or hawks while attacking these large trespassers.

Plate 53

RED-WINGED BLACKBIRD

Common Grackle: *Quiscalus quiscula*

ETYMOLOGY

Quiscalus and *quiscula*: both refer to quail for unknown reasons

Description   A long-tailed, long-billed bird with a pale yellow eye and dark, iridescent plumage that in good light projects purple, blue, bronze, and green. By late May, plain-brown juveniles start appearing alongside their multihued parents.

Size   At just over a foot long, the common grackle is a few inches longer than American robin and northern mockingbird but about 5 inches shorter than American crow.

Common Locations   The trees, gardens, and expansive lawns of the National Mall and the National Zoo draw many nesting pairs of common grackles, as do backyards and parks across the city. Usually scarce in the city during winter, by mid- to late February common grackles start to return to their nesting areas. By late March, the common grackle is once again one of the city's more common birds.

Notes of Interest   The brown-headed cowbird lives in many of the same places as the grackle. Also a member of the New World blackbird family, this bird's nesting habits are very different. Female cowbirds lay their eggs in other birds' nests, leaving their young to be raised by surrogates. Male brown-headed cowbirds have satiny black bodies and brown heads; females are grayish brown. Cowbirds have shorter bills than grackles.

Ecological Role   Much of the common grackle's food is collected while foraging on the ground. Common grackles eat waste grain, including corn. When available, acorns are an important food. About a quarter of a grackle's diet may be animal matter, including grubs, caterpillars, spiders, sometimes smaller birds, eggs, fish, or meat scraps found in garbage.

KEY POINTS

- Common grackles nest in loose groups, often in stands of pines or other conifers. When migrants such as warblers and orioles are just arriving, grackle nesting is already well under way.
- Common grackles spend their falls and winters as members of large, often mixed flocks, feeding with cowbirds, starlings, and red-winged blackbirds on farm fields.
- The larger boat-tailed grackle is found in salt marshes several hours' drive to the east, along the Atlantic coastline. It is not found in the city.

Plate 54

COMMON GRACKLE

European Starling: *Sturnus vulgaris*

ETYMOLOGY

Sturnus: refers to starling; *vulgaris*: common or familiar

Description   The striking starling wears two wardrobes—iridescent, green-bodied, purple-headed plumage late winter to August and then frosty headed but otherwise dark plumage stippled with white triangles the rest of the year. In spring, the bird's long, pointy bill is bright yellow; by late summer, it has changed to blackish. Recently fledged birds are all tan. Compared with blackbirds, starlings are rather short tailed.

Size   European starlings measure about 8.5 inches long, the same length as a gray catbird and at least 2 inches longer than a house sparrow.

Common Locations   European starlings are a common sight around trashcans and food vendors all along the National Mall and at other grassy, park-like settings. They are also common along the city's streets and in its neighborhoods. When they are not nesting, starlings roost in large numbers, spattering trees and buildings with smelly, white droppings. They nest in a wide variety of cavities, including light poles.

Notes of Interest   In 1890 and 1891, about 100 starlings were introduced into New York's Central Park. That population spread across the United States and Canada and into northern Mexico and is now estimated at more than 200 million birds.

Ecological Role   Usually seen moving around in flocks in areas with short grass, European starlings eat seeds, berries, and a wide variety of invertebrates, from millipedes, snails, earthworms, spiders to insects. Starlings find soil invertebrates, such as beetle grubs, while probing and opening small gaps in the turf. They eat garbage and suet at feeders.

KEY POINTS

- Starlings belong to the same Old World family as mynas. Their song is a jumble of gurgles, squeaks, and whistles that may include imitations of other birds' songs.
- Starlings are tough competitors for nest sites, usually prevailing over native cavity-nesting species, such as bluebirds.
- By late May or mid-June, all-tan juvenile starlings roam the city in small to large flocks.

Plate 55

From left: EUROPEAN STARLING & HOUSE SPARROW

House Sparrow: *Passer domesticus*

ETYMOLOGY

Passer: small bird; *domesticus*: belonging to the home or house

Description   The chubby house sparrow is an attractive little bird seen daily by city dwellers. From spring until September, the male has a rich chestnut-colored nape, contrasting white cheeks, a black throat and breast, and a gray crown. On his chestnut wings, the male has a bold white bar. From September to late winter, the male's markings become somber, though the pattern remains. The female is easier to confuse with true sparrows. (House sparrows are weaver finches, as mentioned below.) She has a streaked back, a dingy grayish brown wash below, and a horn-colored bill.

Size   At 6.25 inches long, this familiar bird sets the standard for songbird size comparisons.

Common Locations   House sparrows are found throughout the city and are nowhere more common than along the streets, snapping up crumbs at cafes, stuffing nests into crevices in buildings or even between the lettering of store signs. Plenty of house sparrows live on the National Mall, especially near concession stands.

Notes of Interest   The house sparrow is not a true sparrow but a member of the weaver finch family, whose members are found in Eurasia and in Africa. The female's coloration is similar to that of the native sparrows, such as white-throated and song sparrows.

Ecological Role   House sparrows, like brown rats and street pigeons, are considered "commensal" with humans. This bird thrives in cities and at farm feed lots, where discarded food, seeds, and insects abound.

KEY POINTS

- The chirping of house sparrows is so commonplace that it often goes unnoticed.
- House sparrow pairs construct messy nests of grasses, twigs, pieces of plastic bags, and other materials and often line their creation with feathers. The nest is usually crammed into an enclosed area, such as a bluebird box, but house sparrows sometimes build nests on open branches. In these cases, the nest is globe shaped.
- Introduced to New York City in 1851 and 1852, the house sparrow now occupies all of North America, where it competes for nest cavities with native bluebirds, tree swallows, and others.

House sparrow

Dark-eyed Junco: *Junco hyemalis*

ETYMOLOGY

Junco: Latin for reed or reed-like plant and the somewhat similar reed bunting of Europe; *hyemalis*, refers to winter

Description   This petite sparrow is much more dapper than its other eastern brethren. Instead of brown and streaks, the junco sports a slate-colored head and breast and a whitish bill. The male continues this theme on his back and sides, while the female has browner tones there. Both sexes have white bellies, blackish tails, and flashy white outer tail feathers. The birds make quick chipping notes when they take off.

Size   Averaging 6.25 inches long, the dark-eyed junco is the same length as a house sparrow, although with a shorter body and somewhat longer tail.

Common Locations   In winter and during migration, dark-eyed juncos are found in small flocks in open and edge habitats, such as the brush at the edge of deciduous woods or in backyard gardens. They are found around the National Mall, in Rock Creek Park, Anacostia Park, the National Arboretum, and many other sites from late September through the beginning of May.

Notes of Interest   The "slate-colored" form is found in the East, but five other forms with different coloration are found in different parts of the West.

Ecological Role   As in many songbirds, young and adult juncos add insects to their diets during the summer. In Washington, D.C., during the cold months, they primarily eat grass and weed seeds and birdseed.

KEY POINTS

- Watch for juncos on the ground, or near the ground in shrubs. During migration, they are sometimes seen up in trees singing their trilling song.
- On their breeding grounds to the north and in the Appalachians west of the city, juncos nest in open coniferous forests or forests with both conifers and deciduous trees.

Plate 56

DARK-EYED JUNCO

Northern Cardinal: *Cardinalis cardinalis*

ETYMOLOGY

Cardinalis: refers to the costume of a Roman Catholic cardinal—red robe and high-peaked hat

Description   The northern cardinal is one of the city's most colorful and familiar birds. There are other red birds in the woods—the scarlet tanager, high up in the tree canopy in spring and summer and, rarely within the city, the summer tanager. But unlike these birds, the cardinal is around all year, and it sings out in the open, flits in and out of shrubs, and hops on and below bird feeders. And male and female both have red crests and thick, orange-red bills. The female is mostly a rich buffy tan, with red wings, tail, and crest. Immature birds are colored like females but have blackish bills and lack the reddish on the crest.

Size   Cardinals measure up to 9 inches long. They are about the size of a catbird and a bit shorter than both mockingbirds and American robins.

Common Locations   Look and listen for cardinals just about anywhere in the city where large shrubs and trees grow. They also frequent bird feeders stocked with sunflower seed.

Notes of Interest   There are three members of the *Cardinalis* genus: the northern cardinal, the pyrrhuloxia of the Southwest and Mexico, and the vermilion cardinal of coastal Venezuela and Colombia.

Ecological Role   The cardinal's heavy bill is ideal for cracking tough coverings such as those protecting sunflower seeds. But cardinals do not limit themselves to seeds. They eat berries and invertebrates, from beetles to grasshoppers to caterpillars, spiders, centipedes, and others.

KEY POINTS

- Throughout the year, listen for the cardinal's piercing chip.
- Starting in late winter and continuing well into summer, cardinals sing a loud, rich "*Chewa-CHEER, Chewa-CHEER, Chewa-CHEER, WIT, WIT, WIT, WIT, WIT, WIT.*"
- Cardinals move locally and gather in flocks after nesting season, but they do not migrate long distances.

Plate 57

NORTHERN CARDINAL

House Finch: *Carpodacus mexicanus*

ETYMOLOGY

Carpodacus: fruit biter; *mexicanus*: from Mexico

Description   Both male and female house finches are sparrow-like birds marked with brown streaks below. The male's eyebrow, throat, breast, and rump are colored a flat cherry red. This bird's bill is thick, rounded, and short. The backs of both male and female are dull brown with subtle streaks.

Size   At 6 inches long, the house finch is about the size of the house sparrow, but not as chunky.

Common Locations   House finches live year round in Washington. During nesting season, they often nest in conifers, ivy, and hanging baskets outside of houses and in gardens, such as the Enid A. Haupt Garden behind the Smithsonian Castle.

Notes of Interest   The purple finch is an uncommon, sporadic winter visitor to feeders and parks, but pushes through in numbers in April. Like the house finch, the male purple finch appears doused in a raspberry wash. Unlike the house finch, though, the male lacks brown streaking below, and its coloration is closer to pink than to the house finch's red. Unlike the female house finch, which has a plain brown head, the female purple finch appears masked, with a whitish line above and below a dark eye stripe. Females of both species are otherwise heavily streaked in brown below, and both have streaked brown backs.

Ecological Role   House finches eat seeds, berries, and the buds of trees and shrubs. They sometimes eat small insects.

KEY POINTS

- The house finch is not native to the East. It hails from the arid western states and is native to much of Mexico. Released pet store birds established the eastern population, which in recent decades has spread, reaching their western compatriots.

- The rollicking, rollercoaster song of the male house finch is a familiar spring and summer sound in Washington's street-side trees. After a bit of searching, the male can usually be found, often on a branch of a midsized tree.

- House finches are common visitors to feeders in backyards and at nature centers. In late fall through early spring, they may be found in flocks in weedy areas.

Plate 58

HOUSE FINCH

American Goldfinch: *Carduelis tristis*

ETYMOLOGY

Carduelis: refers to Latin for goldfinch and thistle, an important food; *tristis*, sad

Description   Many first-time viewers think the male American goldfinch is an escaped canary. Almost impossibly bright yellow on head and body, the male is also adorned with a black forehead and black, white-striped wings. The female is yellowish with blackish wings and whitish wing bars. The short, notched tail is black infused with white. In fall and winter, male and female look washed out. Females are mostly grayish, while males still show some yellow on the head and shoulder.

Size   At 5 inches, the American goldfinch is noticeably smaller than house finches and most sparrows.

Common Locations   American goldfinches visit bird feeders, particularly those offering thistle. Left to their own devices, though, they frequent unmown fields, forest edge, and other areas where wild thistles and other seed-producing plants grow in profusion. Usually these habitats have scattered trees and shrubs, which provide both shelter and nesting sites. Watch for them at Kenilworth Aquatic Gardens, the C&O Canal National Historical Park, and in flower gardens, including the extensive plantings at the National Arboretum and National Zoo.

Notes of Interest   The pine siskin, a sporadic winter visitor to the city, sometimes appears with winter goldfinch flocks. This bird has the same build as the goldfinch but is heavily streaked with brown. And only the male shows a prominent yellow wing bar. Both siskins and goldfinches have thin, pointy tipped bills.

Ecological Role   Goldfinches are primarily seedeaters, feeding in gardens on coneflowers, black-eyed Susans, salvias that have gone to seed, sunflowers, thistles, and the seeds of some trees such as elms. They also eat insects.

KEY POINTS

- In late spring and summer, male goldfinches perform a bouncy roller-coaster flight while calling *perCHICKaree* or *potATOchip*. The goldfinch's song is more complex, a sweet, fast jumble.
- American goldfinches frequently nest in small street trees, the female placing the tightly wound cup nest in the fork of a branch, usually lower than 20 feet above the ground.

Plate 59

AMERICAN GOLDFINCH

More D.C. Birds

During a visit to Washington, D.C., you can expect to spot many more birds than those described on the previous pages. I was limited for space, so only chose 38 birds that would be among the easiest to see, depending on the season. If you feel a strong pull toward birding, you will find that the city holds many avian surprises for you. The more you look, the more you will see. The *Official List of the Birds of the District of Columbia*, revised in 2012 by the Maryland / District of Columbia Records Committee of the Maryland Ornithological Society, includes 331 species. (Two of these species no longer exist in the area—ruffed grouse and greater prairie-chicken. One is extinct: passenger pigeon.) The surrounding and far larger states of Maryland and Virginia have larger bird lists of 444 and about 470, respectively. Washington, D.C., is a top city-birding destination for several reasons, including its location on the Fall Line and between north and south, its two rivers, its temperate climate, and its proximity to the country's largest estuary, the Chesapeake Bay.

It's a perfect place to conduct a big year. This is a bird-watching challenge in which a birder tallies all the bird species he or she sees (or hears, depending on how he or she wants to tally) in a calendar year. In 2011, for example, Washington, D.C., resident Jason Berry undertook a big year, targeting different birding hot spots at different times to maximize the numbers of species he saw. He chased rare birds his friends and birding hotlines reported. And he did much of this on foot and by bicycle, all within the city limits.

By year's end, Berry had seen 218 of the city's 331 recorded species, the most tallied in a year by anyone since 1983. Berry enters data on bird species he finds on each bird walk on a citizen science database called eBird (www.ebird.org). Such reporting adds to the knowledge of the city's biodiversity. This kind of bird distribution information helps park managers set conservation priorities. It turns enjoyable hours pursuing an outdoor hobby into valuable information. In recent years, similar volunteer monitoring projects have been undertaken for butterfly, dragonfly, amphibian, and other populations. Will Berry undertake another grueling big year within the city? "Not anytime soon," he said, "but maybe a big day."

What are some of the other birds you might see in Washington, D.C., through the seasons? Here I list some species you can expect, with best time of year indicated where necessary.

Potomac River locations and areas such as the Tidal Basin and Constitution Gardens host many waterfowl in fall, winter, and the first days of spring. Often small flocks but sometimes large gatherings can be found. In addition to the species described on previous pages, look for gadwall, green-winged and blue-winged teal, lesser scaup, bufflehead, and ruddy duck, among others. The somewhat duck-like pied-billed grebe often shows up on open water during migration.

In the skies over the city and its parks, turkey vultures and black vultures put in regular appearances. The turkey vulture has a red head and dark, two-toned wings. It holds its wings in a shallow "V," while the black vulture's wing profile is more or less flat. The black vulture also sports white patches on its primary flight feathers.

A rare sight in the 1970s, bald eagles are now often seen in the city, particularly along the Potomac and Anacostia Rivers. Each year, a pair or two nests in the city. The eagle's rebound is a wildlife success story that followed the ban of insecticide DDT and the strengthening of conservation laws. In 2011, Virginia's nesting eagle population was more than 730 pairs, according to the Center for Conservation Biology. The Maryland nesting population is likely more than 500 pairs, according to the Maryland Department of Natural Resources.

Bird-hunting accipiters frequent various D.C. wild places, especially from fall to early spring. Separating the sharp-shinned hawk and somewhat more plentiful and larger Cooper's hawk is a challenge. Note if the tail is flat ended (sharp-shinned) or rounded (Cooper's) and if the flight profile of a soaring bird is cross-like, as in Cooper's, or somewhat "T"-like due to the sharp-shinned's shorter-headed profile.

Peregrine falcons nest on some buildings and bridges in the area, and they and their smaller relatives the merlin and American kestrel are regularly seen during migration at places such as Hains Point.

During April, Caspian terns and Bonaparte's gulls appear on the Potomac River, en route to nesting areas well to the north. Forster's terns are also seen along the river but usually from summer into fall.

Two sandpipers—spotted and solitary—are often seen along the muddy shores of the Potomac, Anacostia, and Rock Creek. Less often seen but present each migration are greater and lesser yellowlegs and other species more commonly seen along the coast. The stripy, football-shaped Wilson's snipe is a winter visitor to muddy, grassy spots, often staying hidden until accidentally flushed in such spots as Kenilworth Aquatic Gardens or Anacostia Park. Grassy fields and park playing fields are habitat for the killdeer, a large two-banded plover often found away from water.

Yellow-billed cuckoos nest and migrate through Rock Creek Park and other mature forests in the city. The similar black-billed cuckoo passes through in migration.

Eastern screech-owls can be heard early in the a.m. at spots in Rock Creek Park and elsewhere. They are sometimes seen roosting in wood duck boxes. The larger barred owl roosts and nests in mature woods such as those along the C&O Canal, and at Glover-Archbold and Rock Creek parks, and the great horned owl, the largest of the region's owls, may be present but is rarely seen.

The common nighthawk is a special migration bird in Washington, D.C. Late August to mid-September is the best time to watch for them at dusk around the Washington Monument's lights and over the National Mall, over the U.S. Capitol, as well as other spots in the city. They do migrate through in May as well. Watch for the white bands on the wings and the tilting flight.

The region's only common hummingbird, the ruby-throated hummingbird, may be seen in gardens, such as those alongside the Smithsonian museums, at Dumbarton Oaks, and at the Smithsonian's National Zoo. In spring, tuliptree blossoms and in summer trumpet vines are among the native plants attracting them. Rarely, a rufous hummingbird or some other western species may stray to the city and linger around a feeder. This usually happens November or later, after the ruby-throated hummingbirds migrate south.

The eastern phoebe, a tail-pumping gray flycatcher, is a typical early spring arrival, staying to nest through summer and lingering well into fall. It is very similar in appearance to another common summer bird, the eastern wood-pewee, but this bird has wing bars, and a different song (pee-weeee!). It is more often found high in the trees, not down low like the phoebe. The great crested flycatcher nests in forested parts of the city and is identified by its shaggy crest, rusty wings and tail, and yellow belly. The "pit-zee" of the drab Acadian flycatcher is also a familiar sound in D.C. woodlands such as Rock Creek Park, Kenilworth Aquatic Gardens, and Glover-Archbold Park.

Over grassy fields such as the National Mall or over water along the Potomac and Anacostia Rivers, watch for barn swallows, tree swallows, and northern rough-winged swallows, all of which nest in the area. During migration, purple martins, bank swallows, and cliff swallows may pass through the city as well.

In Washington, D.C., there are two crow species, virtually identical by sight but identifiable by voice much of the year. The Amer-

Blue Jay

ican crow utters a familiar *caw caw caw*, while the fish crow, a bird usually near water, calls a nasal *cuh-uh*. From April into fall, however, young birds of both species utter a variety of vocalizations that can make voice identification tricky.

At least 36 wood warblers have been sighted in Washington, D.C., most during migration. The previously described yellow-rumped warbler is the only one regularly seen in winter. A few species nest within the city, singing well into summer. These include the seldom-seen and often-heard northern parula (its song a quickening and ascending "*z-z-z-z-zzzyUP!*"), the pine warbler (which nests at the National Arboretum), the Louisiana waterthrush (Rock Creek Park and the C&O Canal), and the ovenbird (Rock Creek Park and Fort Dupont Park). In addition, a few pairs of the striking gold prothonotary warbler usually occur along the river at Kenilworth Aquatic Gardens and along the C&O Canal in D.C., where yellow-throated warbler also nests.

Other common treetop denizens during summer include the red-eyed vireo and boldly colored scarlet tanager. In open areas with large trees, the warbling vireo, a drab, often overlooked bird sings its winding song. The vireos and tanager usually take a special, neck-craning effort to see, but their songs are easy to hear, as is the wheezy chatter of the tiny blue-gray gnatcatcher.

The wood thrush, the official bird of the nation's capital, can frequently be heard and sometimes seen in the forested parks of the city. Its song is flute-like and rising at the end. Another thrush, the veery, has a spiraling song, and can be heard in some forested parks in the city. The hermit thrush, a quiet unobtrusive bird with a rusty tail, is the only brown-backed thrush to winter in Washington. The Swainson's thrush, with its buffy eye ring, is a familiar sight in the woods during migration periods. Also, various parks have nest boxes put out for eastern bluebirds. Watch for these boxes, especially in early spring, and you may be rewarded with views of one of the region's most spectacular birds.

The eastern towhee sings *drink-your-tea* from woodland edge and forest undergrowth and is found in the National Arboretum's azalea gardens and at Fort Dupont Park, among other areas. One of the city's most colorful birds is the indigo bunting. It nests in some parts of the city where forest edge meets open field. The all-blue male sings from high perches and is hard to see well without binoculars. The meadow by the old Capitol columns at the National Arboretum, open areas at Roosevelt Island, Fort Dupont Park, Rock Creek Park, and the C&O Canal are good places to seek this blue beauty.

Early spring, fall, and winter is sparrow time in Washington, D.C. The familiar song sparrow is around year round as is the house sparrow (which is not a sparrow at all). Visitors include the white-throated sparrow, one of the commonest winter birds, along with the possibility of seeing swamp, chipping, fox, and white-crowned sparrows.

Northern Raccoon: *Procyon lotor*

ETYMOLOGY

Procyon: before dog; *lotor*: washer

Description   Washington, D.C., has a large raccoon population, but most people see them only at night, often near trashcans. The bushy, ringed tail, the white-outlined black mask, and the somewhat pointy snout distinguish the northern raccoon from cats and dogs. The coat is brown grizzled with grayish or blackish.

Size   From nose to ringed tail tip, raccoons measure between 2 and 3 feet long. Size varies by region and the abundance of local food supplies. Particularly well-fed individuals may weigh nearly 50 pounds, although this is an extreme. Some weigh less than 15 pounds.

Common Locations   Raccoons inhabit much of the city, sleeping (even mating) in trees but descending at night to forage in and along streams and other wet areas and to scavenge in neighborhoods. In residential areas, they den in large trees with hollows.

Notes of Interest   The name raccoon originates from the Algonquian language and means "scratching with its hands." Many people believe raccoons wash their food, but this is likely just the animals manipulating food items to distinguish edible from inedible parts.

Ecological Role   A widespread omnivore willing to exploit many feeding possibilities, the raccoon eats nuts and berries, trash and carrion, cicadas, crayfish, fish, frogs, snakes, bird eggs and nestlings, and rodents.

KEY POINTS

- Raccoons lead solitary lives, except when males linger near females during mating season and from spring to late summer, when young raccoons still need their mothers' care.
- Despite its apparent girth, the raccoon is nimble in trees and can descend head first with the aid of its hind feet, which rotate 180 degrees.
- Small hand-like prints in mud by the water's edge provide evidence of raccoons' nightly sojourns.

Plate 60.

NORTHERN RACCOON

Eastern Chipmunk: *Tamias striatus*

ETYMOLOGY

Tamias: collector or keeper; *striatus*: striped

Description   "Aaaaw" is usually the reaction when a chipmunk scurries past a visitor to the National Zoo or other wooded D.C. park. Just a few ounces but with tons of appeal, this energetic, fist-sized rodent is much smaller and less bushy tailed than an eastern gray squirrel. The eastern chipmunk has orangish hindquarters and on each side a prominent white stripe bordered with thinner black stripes, plus a stripe down over the spine. The furry gray tail is about as long as the body.

Size   Eastern chipmunks measure between eight and a half inches and almost a foot long, including the tail. They weigh between 2.5 and 5 ounces.

Common Locations   Two places where eastern chipmunks are easy to see are Rock Creek Park and the adjacent National Zoo. They frequent forest and forest edge where there are plenty of fallen logs, rock piles, or other cover.

Notes of Interest   The eastern chipmunk is the largest of North America's 22 chipmunk species, and the only one found in the East.

Ecological Role   The eastern chipmunk does much of its foraging on the ground, but also climbs into shrubs and trees such as oaks when food is available there. When available, acorns, beechnuts, hickory nuts, and walnuts are important foods. But the chipmunk diet also includes various seeds, corn, wild grape, mulberry, insects such as cicadas, slugs and snails, birdseed, and sometimes small amounts of carrion.

KEY POINTS

- Unlike eastern gray squirrels, eastern chipmunks stuff their cheeks full of food and take this stash back to underground storage, or cache, sites.
- In spring, eastern chipmunks give birth to a litter of three to five young.
- The eastern chipmunk's piercing alarm call is often mistaken for a bird's chip.

Plate 61

EASTERN CHIPMUNK

Eastern Gray Squirrel: *Sciurus carolinensis*

ETYMOLOGY

Sciurus: squirrel; *carolinensis*: of the Carolinas

Description The eastern gray squirrel is usually silver gray above, with a burn of chestnut on its back in spring and summer. Below, its white belly fur provides a sharp contrast. The fluffy gray tail is edged with white. In Washington, D.C., and its suburbs, however, you will also find melanistic (black) individuals.

Size Including its long, bushy tail the eastern gray squirrel measures about 18 inches long and weighs between 14 and 25 ounces.

Common Locations Trashcans along the National Mall and at the National Zoo are among the many places you will find these bold animals. Although often seen on the ground, eastern gray squirrels spend much of their time feeding, resting, and nesting in trees. All the city's parks and tree-lined streets provide gray squirrel habitat.

Notes of Interest Black-phase gray squirrels naturally occur much farther north than Washington, but you will see them in the city because they were introduced at the National Zoo in 1902 and 1906. The released squirrels (18 individuals) hailed from Ontario, the Canadian capital. Squirrels of this color phase now live in and beyond the city, having reached the Maryland and Virginia suburbs.

Ecological Role Eastern gray squirrels bury some acorns and other nuts, attempting to relocate them later when needed. Some of these are forgotten and eventually grow into trees. The eastern gray squirrel's diet varies by season: Acorns, hickory nuts, beechnuts, and walnuts are important when available. In spring, eastern grays eat tree buds, flowers, and seeds. In summer and early fall, berries and other fruits go on the menu along with some insects and the occasional egg or baby bird. They also frequent bird feeders, which in winter supplement their diet of cached tree nuts and other meager pickings. These squirrels sometimes eat tree bark as well.

KEY POINTS

- Eastern gray squirrels build bulky nests in tree branches, which they fashion out of leaves and twigs.
- Many female eastern gray squirrels raise two litters a year: one in spring and one in late summer. Reproductive success often depends on the bounty of the year's nut crop.

Plate 62

EASTERN GRAY SQUIRREL

- Unlike the red squirrels found to the north and in the
  mountains to the west, eastern gray squirrels do not defend
  territories. However, you may see dominant individuals
  chase other squirrels from feeding spots.

Woodchuck (Groundhog): *Marmota monax*

ETYMOLOGY

Marmota: marmot; *monax*: digger

Description For a rodent, the woodchuck is very large and lumbering, justifying its other common name, groundhog. This brown-coated mammal has a long, thick, rounded tail, small ears, and a large head.

Size Woodchucks vary in length from about one and a half to a bit less than 3 feet long. They weigh between 5 and 14 pounds.

Common Locations Woodchucks favor open grassy areas, often near forest edge. There they spend their days feeding on grasses, clovers, and weedy plants. The George Washington Memorial Parkway and open areas of the Rock Creek Parkway may yield roadside sightings. Grassy, open stretches at the National Arboretum, Kenilworth Aquatic Gardens, and the C&O Canal are other places to watch for this animal.

Notes of Interest One of North America's largest rodents, the woodchuck is a member of the squirrel family.

Ecological Role Woodchucks damage gardens and crops, but they are also beneficial. Their droppings fertilize the soil, and their digging aerates and churns it. Rabbits, raccoons, and foxes may use woodchuck burrows after the rodents move out. Coyotes, foxes, and red-tailed hawks prey on groundhogs.

KEY POINTS

- During winter hibernation, a woodchuck's heartbeat drops from about 100 beats per minute to 4.
- February 2 is Groundhog Day, a U.S. and Canadian holiday when selected woodchucks are supposed to leave their burrows and seek their shadows to determine how much longer winter will continue.
- The woodchuck is the only eastern marmot. Five other species inhabit the West.

Plate 63

WOODCHUCK, OR GROUNDHOG

Eastern Cottontail: *Sylvilagus floridanus*

ETYMOLOGY

Sylvilagus: wood rabbit; *floridanus*: of Florida

Description   The eastern cottontail is the city's only wild rabbit. Its coat is brown or grayish brown. It has long ears, a white belly, and its tail has a cottony white underside.

Size   Weighing in at 2 to 4 pounds, an eastern cottontail rarely grows beyond 18 inches long.

Common Locations   Watch for eastern cottontails in open areas with plenty of nearby cover such as overgrown fields or woodland edge. They may show up in backyards and around large gardens. You are likely to find a few at Rock Creek Park, the National Arboretum, Kenilworth Aquatic Gardens, and some other areas.

Notes of Interest   The only wild rabbit found over much of the eastern United States, the eastern cottontail is very fecund but usually short-lived.

Ecological Role   Cottontails are herbivores feeding on a wide variety of plants, including clover on lawns, sapling twigs, and shrub and tree buds in winter. Foxes, red-tailed hawks, coyotes, and other predators hunt them.

KEY POINTS

- A female eastern cottontail may live most of her life within a five-acre area, depending on the quality of habitat. Males range more widely.

- Eastern cottontails are mostly nocturnal, but you may see them at dawn and at dusk.

- Most of us see eastern cottontails making short hops but when they urgently need to reach cover, they can bound at least several feet at a time.

Plate 64

EASTERN COTTONTAIL

White-tailed Deer: *Odocoileus virginianus*

ETYMOLOGY

Odocoileus: from Greek for "hollow tooth"; *virginianus*: of Virginia

Description By far, this is the city's largest wild mammal. In spring and summer, the white-tailed deer's coat is reddish brown. By winter, the fur color has changed to grayish brown. Males have antlers, which they shed each winter and regrow during the summer. The tail is long for a deer, tan above and white below, with white fringing on the sides. Both male and female have white throats and bellies. Fawns are spotted with white.

Size White-tailed deer may reach almost 4 feet tall at the shoulder. They grow up to 7 feet long, measured nose to tail tip. Males weigh up to 300 pounds; females rarely reach 200.

Common Locations White-tailed deer like patchworks of forest and clearings, exactly the landscape created when forested parks abut residential areas. Watch for them early or late in the day at most parks, including Glover-Archbold Park, Kenilworth Aquatic Gardens, Rock Creek Park, and the C&O Canal National Historical Park.

Notes of Interest Deforestation and overhunting greatly reduced deer herds until the 1940s, when re-stocking and reforestation began to bring them back to the area. Today, local deer populations are booming.

Ecological Role White-tailed deer graze and browse the forest and adjacent open areas, including yards. A high-density deer population reduces cover needed by ground-nesting birds, other mammals, reptiles, amphibians, and many invertebrates. It also cuts back on floral diversity. Deer have to eat a lot of vegetation each day, and a large population takes a toll on tender forbs and shrubs and tree saplings. Coyotes are new arrivals in the city. On occasion, they might kill a young or injured deer, but this canid can never replace the deer's chief predators, wolves and pumas. The area's wolves and pumas, however, are long gone, wiped out by hunting well over a century ago. In the city and its suburbs, the car is now the deer's chief "predator."

KEY POINTS

- Watch for the split hoof tracks of white-tailed deer along trails and in moist open areas. A track resembles two large, 2- to 3-inch teardrops, their tapered ends facing forward.
- Although often seen during the day, white-tailed deer do much of their feeding and moving around at night.

Plate 65

WHITE-TAILED DEER

- Most herds people see consist of a doe with her young. Bucks travel around in small groups much of the year. Males and females may be found together in winter herds.

More D.C. Mammals

A naturalist typically sees far more birds than mammals in Washington, D.C., unless of course you count eastern gray squirrels, humans, and their pets. Many other mammals inhabit the city. Your best chance to see them is at dawn and dusk, or after dark. In the dark of night, for example, opossums, North America's only marsupials, prowl parks and yards, and southern flying squirrel glide from tree to tree in forested areas.

Two introduced mammals have long been familiar Washington, D.C., residents. More often heard than seen, the city's brown or Norway rat population is impressive. At night, rats are sometimes seen along the city's sidewalks and streets, mistaken at first for thin-tailed nocturnal squirrels. And it is likely that most of the city's houses host house mice. In the woods and fields, native small mammals often present themselves as small blurs flashing across paths. These might include northern short-tailed shrews; woodland, meadow, or southern red-backed voles; and North American or white-footed deermice. Watch your step, a raised tunnel across or along the path might be the handiwork of an eastern mole.

Along the Potomac and Anacostia Rivers and their associated wetlands, the beaver population has rebounded. Beaver families fell trees and dam streams, creating ponds that provide habitat for varied wetland wildlife. Watch for evidence of whittled-down young trees at the C&O Canal, Kenilworth Aquatic Gardens, and other riverside spots. Beavers fell these trees not only to use them for building dams and lodges but also as important winter food. Much smaller muskrats live in some of the city's marshes, including Kenilworth Aquatic Gardens. Watch for their dens, mounds of water plants cut, carried, and piled there by these busy rodents. A lucky early or late visitor to the Anacostia or Potomac River marshes might spy mink or river otters, slinky predators that, while rarely seen, have had some rebound in numbers within the region.

With all the connected parklands, it should be no surprise that some large mammals live within the city limits. A recent and now entrenched new resident is the coyote, which has been seen in Rock Creek Park and adjacent residential neighborhoods at night. Camera traps within Rock Creek Park snapped photos of coyotes at a road-killed deer carcass. The red fox is more often seen. These orange-furred canids often hunt at dawn or dusk in fields adjacent to forests.

Northern Raccoon

The more stealthy gray fox is seldom seen but is present in some parks.

At twilight and after dark on warm nights, several bat species wheel over the city or its parks. A survey conducted by the District Department of the Environment in 2010 recorded these five bat species: silver-haired, big brown, eastern red, and evening bats, and eastern pipistrelle. In addition, the survey recorded bats unidentified to species but within the genus *Myotis* (mouse-eared bats).

CHAPTER

5

plants

AQUATIC PLANTS

Common Cattail: *Typha latifolia*

ETYMOLOGY
Typha: bulrush; *latifolia*: broad-leaved

PLACE OF ORIGIN
Native

Description   From spring into early summer, the namesake brown female flowerheads resemble straight corndogs. They grow up to a foot and a half long and about an inch wide. The male flowers are yellow and sit above the female flowerheads, but they quickly wither after releasing their pollen. The mature female seedhead eventually goes from firm and brown to loose with patches of wispy white, as flowers yield to fluffy, wind-dispersed tiny seeds. The long leaves are narrow, only up to an inch wide. They resemble green sword blades. Stiff, dead stems remain standing through the winter, rattling in the breeze. New shoots emerge in early spring.

Size   The common cattail grows up to 10 feet tall.

Common Locations   Cattail thrives in wet shallow areas, including those along the Potomac and Anacostia Rivers. Watch for it along the C&O Canal and at Kenilworth Aquatic Gardens, and almost anywhere there are roadside ditches or small ponds.

Notes of Interest   Common cattail spreads both by seeds and by rhizomes, or underground stems. As rhizomes move into the mud, they promote land building, extending the shoreline inward, gradually reducing a wetland's size.

Ecological Role   Muskrats eat the roots of this plant and use its leaves to make their homes. Birds use the fluff to line their nests. Ducks and geese sometimes eat the seeds, rhizomes, or shoots. Male red-winged blackbirds often sing from atop cattails, while females build their nests between the plants' stalks. Waterfowl and sparrows find shelter among the plants.

KEY POINTS

- Hailed by many as one of North America's most versatile edible wild plants, the common cattail's young shoots and stalks, green flower spikes, sprouts, and roots are edible. Even its pollen can be used to make flour.
- A seedhead contains up to 250,000 seeds.
- Cattail can crowd out other native plants but also helps control erosion and soaks up pollutants.

Plate 66

COMMON CATTAIL

Common Reed: *Phragmites australis*

ETYMOLOGY
Phragmites: refers to fence; *australis*: southern

PLACE OF ORIGIN

European form dominates native form

Description   The plumes adorning the top of the common reed consist of many tiny pale flowers. Its blue-green leaves grow on long stems resembling tall, upright cornstalks. Leaves drop in fall, but stalks and plumes remain through the winter. In spring, the plant regrows in lush green. This plant spreads quickly via its rhizomes or underground stems.

Size   Common reed is a large perennial grass growing up to 15 feet tall.

Common Locations   You will see this plant along stretches of the Anacostia River and in other wetland areas in and around the city.

Notes of Interest   In other parts of the world, the common reed is used for thatch, to build boats, and to feed livestock. Native Americans used it to make arrow shafts, musical instruments as well as mats and other household items.

Ecological Role   Common reed provides food and cover for some wetland wildlife but crowds out native species, including cattails. Its dense stands may filter out pollutants and prevent erosion, but they also can change wetland hydrology and diversity and make wet areas more fire prone.

KEY POINTS

- Common reed is one of the most widespread plants on Earth.
- While many consider it an introduced weed, scientists have found that there is a native form. The European form, however, is more robust and has overtaken the native form in many parts of the East, especially in disturbed soils.

Plate 67

COMMON REED

Red and White Clover: *Trifolium* spp.

ETYMOLOGY

Trifolium: three leaves

PLACE OF ORIGIN

Europe

Description   Red clover is usually taller and has larger leaves and flowerheads than white clover. The tiny, pea-like flowers grow in rounded heads that are up to 1.5 inches wide in the red and up to 1 inch wide in the white. Red clover flowerheads are purplish while those of the white are true to their name. The trifoliate leaves of the red are much larger than the dainty, more rounded white clover leaves and often show a pale white "V" on each leaflet. The white has a much paler, U-shaped chevron on each leaflet.

Size   Red clover grows up to 20 inches tall; white clover rarely reaches 10 inches tall.

Common Locations   Look for red clover in fields and disturbed soils, such as those along roadsides. On lawns and park playing fields it should be easy to find white clover. These clovers flower May through October.

Notes of Interest   Clovers are legumes, members of a plant group that includes peas, beans, alfalfa, and redbud. Like many other legumes, these plants "fix" nitrogen in the soil. Nodules on clover roots contain bacteria that transform nitrogen into a form that naturally fertilizes the soil.

Ecological Role   The larger red clover attracts longer-tongued bumblebees. White clovers draw honeybees, as many unfortunate barefooted children have found while romping on lawns. White-tailed deer, cottontail rabbits, and other mammals readily eat these plants, and many butterflies are drawn to clover nectar.

KEY POINTS

- Both red and white clover are native to Europe.
- Red clover is Vermont's state flower.
- Beekeepers recognize white clover as an important food source for commercial honey-producing bees.
- People can eat clover, too. Flowerheads and young leaves, if soaked in salt water or boiled, can be eaten. They are nutritious but not very tasty. There are also recipes for clover fritters.

Plate 68

RED AND WHITE CLOVER

Common Milkweed: *Aesclepias syriaca*

ETYMOLOGY

Aesclepias: refers to the ancient Greek god of medicine and healing, Aesculapius; *syriaca*: from Syria

PLACE OF ORIGIN

Native

Description   A distinctive oval-leafed perennial field plant that releases white, fluffy windborne seeds in autumn, which children try to catch in the air for good luck. Flowering takes place in summer. The small, dark pink blooms sit atop the plant in large clusters. The lance-shaped but rounded (oval) leaves are 3 inches to a foot long, with a prominent central vein. They are grayish and feel fluffy below (covered with soft hairs). The large, ridged seedpods have a bumpy, rough surface.

Size   Common milkweed grows 3 to 5 feet tall.

Common Locations   Fields, forest edges, roadsides, and other sunny, unmown areas provide habitat for common milkweed. The C&O Canal National Historical Park and Rock Creek Park are among the many places you will see this plant.

Notes of Interest   Although the species name refers to Syria, this plant is native to North America.

Ecological Role   Common milkweed is an important host plant for monarch caterpillars and provides nectar to many butterfly species.

KEY POINTS

- This plant's common name comes from the milky substance released when parts of the plant are snapped. The substance is slightly toxic.
- Other milkweed species occur in the area, including butterfly weed and swamp milkweed. Their leaves are smaller or more narrow and the flowers differ.

Plate 69

COMMON MILKWEED

Common Mullein: *Verbascum thapsus*

ETYMOLOGY
Verbascum: from Latin *barba* for beard, referring to hairy leaves; *thapsus*: named after a town in what is now Tunisia

PLACE OF ORIGIN
Eurasia

Description   One of the city's most easily identified plants. Flowering from late spring through summer in sunny, disturbed soils, common mullein is a tall, velvety plant with large, rounded gray-green leaves covered with woolly hairs. A tall club-like spike of densely clustered yellow flowers rises above. The flowers are between three-quarters of an inch and an inch wide.

Size   Common mullein grows up to 6 feet, or sometimes 8 feet, tall. The bottom leaves grow as long as 1 foot.

Common Locations   Common mullein grows in many sunny locations in the city's parks and along road margins.

Notes of Interest   Common mullein is biennial. The first year, leaves form a rosette resembling a large, fluffy cabbage. This remains through the winter. The second year, the flowery spike rises from the center of the rosette of leaves.

Ecological Role   Over winter, common mullein's rosette shelters many insects, including ladybugs. Some songbirds, such as American goldfinches, sparrows, and indigo buntings, eat the seeds.

KEY POINTS

- Common mullein is in the snapdragon family.
- Its fluffy grayish leaves mislead some people into thinking it is the common garden plant known as lamb's ear.
- Over the centuries, European and Native American peoples have used this plant for medicinal and everyday purposes, including using the stems to make wicks and torches and leaves to line shoes in winter.

Plate 70

COMMON MULLEIN

Dandelion: *Taraxacum officinale*

ETYMOLOGY
Taraxacum: Greek for disorder remedy; *officinale*: medicinal
PLACE OF ORIGIN
Eurasia

Description   Originally from Eurasia, the dandelion is now a worldwide fixture in settled, open, sunny areas with at least some rain. Its rangy dark green leaves are strongly notched, forming a repeated arrowhead pattern pointing toward the leaf tip. Yellow flowers wither and yield to fluffy, pincushion-like seedheads packed with hundreds of white-tufted seeds. Wind or a heavy exhalation launch these seeds to new growing sites. The dandelion is among the earliest wildflowers to produce nectar and then seeds. It flowers from late winter through fall. The stalk holding the flower and seed-head is hollow.

Size   The flowers are 1 to 2 inches wide; the plant may stand as tall as 18 inches.

Common Locations   Dandelions grow throughout Washington, D.C., in any patch of lawn, in sun-filled parks, in vacant lots, along sidewalks, in essence wherever the human footprint is heavy. The question is: Where won't you find this plant? Answer: It won't grow in water, can't grow on pavement (but between the cracks), and usually not in full forest shade.

Notes of Interest   Dandelion can regrow from the taproot if plucked and new plants start from the wind-blown seeds.

Ecological Role   The nectar of this nutritious "weed" attracts pollinating bees, butterflies, and other insects and also draws lady beetles. American goldfinches, house finches, chipping sparrows, indigo buntings, and other seed-eating songbirds eat the seeds.

KEY POINTS

* Edible wild-plant aficionados hail the dandelion (when collected from unsprayed areas) as perfect for various dishes, including salads, fritters, and a coffee-like drink made from the roots. The flower buds can be boiled or pickled and the flowers fried after being dipped in batter.
* Dandelion is rich in minerals and vitamins A and C and has various medicinal uses.
* Children and kids at heart help spread this plant by plucking the stem and blowing the parachute-like seeds into the air.

Plate 71

DANDELION

Spotted or Orange Jewelweed: *Impatiens capensis*

ETYMOLOGY

Impatiens: refers to the apparent impatience (rapidity) with which many of the plants in the genus grow and flower; *capensis*: from the Cape of Good Hope (an error in geography, as the plant is native to North America)

PLACE OF ORIGIN

Native

Description   Growing in lush patches, this native annual is a familiar sight during the warmest months along the edges of streams, canals, freshwater marshes, and other wet areas, as well as in moist woods. The leaves are oval with some rough edging. The stalks are smooth and juicy when snapped. Present from summer through midautumn, orange pitcher-shaped flowers hang from delicate stalks and are often spotted with reddish brown. A spur sweeps under each flower like a curled-in tail. The green fruits resemble tiny, ridged sausages. The name jewelweed likely came from the way water beads on the smooth leaves, although some believe the delicate flowers or fruits inspired the name.

Size   Depending on sunlight, moisture, and soil, spotted jewelweed plants grow to between 2 and 5 feet tall.

Common Locations   Jewelweed grows in patches along the edge of the C&O Canal and in various locations along the shores of Rock Creek and the Potomac and Anacostia Rivers.

Notes of Interest   Another jewelweed species, the yellow jewelweed, occurs in the region but is far less common. This species has yellow instead of orange, spotted flowers.

Ecological Role   Hummingbirds probe jewelweed flowers for nectar. Goldfinches, grosbeaks, and other songbirds eat the plant's seeds. Bumblebees and sphinx moths also visit the flowers, and deer and rabbits may nibble the leaves.

KEY POINTS

- Jewelweeds are temperate impatiens, belonging to the same large genus as the showy summer annual impatiens, such as New Guinea impatiens.
- Jewelweed fruits give this plant another common name, touch-me-not. When touched, taut ripe seedpods snap open in a trigger-like action that sprays tiny seeds in different directions.

Plate 72

SPOTTED OR ORANGE JEWELWEED

- Sap from crushed jewelweed leaves is considered a natural salve for the itching caused by poison ivy rashes.

TREES

Conifers

Most conifers are evergreen. They produce cones containing the trees' reproductive structures. Leaves are usually needle-like but in some species are scale-like, as in redcedars.

Eastern White Pine: *Pinus strobus*

Description   Native to the Northeast and Appalachians, the eastern white pine is one of the most commonly planted pines in the city, including around the monuments, and on the White House and U.S. Capitol grounds. It is the East's tallest tree, with a record height of 220 feet tall, although it does not reach near this height in the city. The straight, flexible, 4- to 8-inch-long needles grow in bunches of five. The cones, which sometimes reach 8 inches long, are tipped with sticky white resin and are not pointed as in many other pines' cones. The trunk, in rare cases, can reach up to 4 feet in diameter. Limbs ring the trunk in a wagon-wheel pattern, with broad gaps between branch groupings. Young bark is smooth and silvery; bark on older trees is darker and rough.

Ecological Role   Valued in the construction industry, eastern white pine also suits the needs of wildlife. Chickadees and other songbirds eat the seeds, as do gray squirrels, mice, and chipmunks. White-tailed deer sometimes browse the branches. Owls roost in groves of white pine and other conifers, and rabbits and other small mammals find refuge beneath the low-spreading branches.

Virginia Pine: *Pinus virginiana*

Description   Also called scrub or Jersey pine, this is a thin-crowned, slim-trunked, midsized tree. The Virginia pine's needles are much shorter than those of the eastern white pine—just one to 3 inches long—and grow in twisted, V-shaped pairs, not groups of five. The cones are much smaller, too, reaching 1.5 to 3 inches, and they are a bit prickly. Virginia pines often grow in stands and rarely live longer than 80 years. This tree has flaky gray-brown bark and usually grows 20 to 40 feet tall, rarely reaching 60 feet. Trunk diameter may reach one and a half feet.

Ecological Role   Used in the pulp paper industry, Virginia pine also benefits wildlife, providing not only shelter but also some food for browsing white-tailed deer. Gray squirrels, chipmunks,

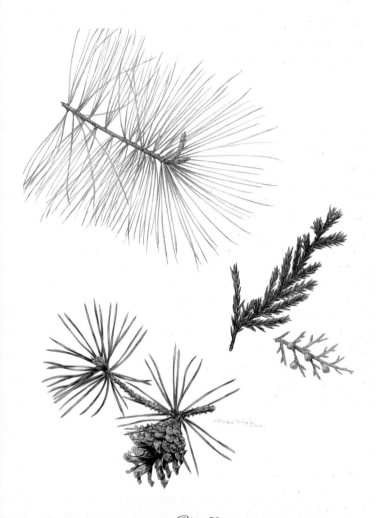

Plate 73

From top: EASTERN WHITE PINE,
EASTERN REDCEDAR & VIRGINIA PINE

and various birds eat the seeds. This is a sun- and heat-tolerant tree that often grows in disturbed soils.

Eastern Redcedar: *Juniperus virginiana*

Description   A native juniper typically found in overgrown fields turning to forest, the eastern redcedar provides a year-round conical accent of green. Like the Virginia pine, this tree rarely reaches 60 feet tall, with a trunk width up to 2 feet in diameter. The bark is gray brown, with longitudinal lines and thin, pealing strips. Young trees have spiky foliage and pointy tops; older ones have more oval crowns and flatter, scale-like leaves. In fall, eastern redcedars become laden with masses of mature frosty-blue, berry-like seed cones.

Ecological Role   Owls roost in redcedars and cedar waxwings and other songbirds dine on their fruit-like cones. Song sparrows, northern mockingbirds, and other songbirds nest in this tree. Eastern redcedar is a host plant for the juniper hairstreak butterfly.

Eastern White Pine

Virginia Pine

Eastern Redcedar

Ailanthus: *Ailanthus altissima*

ETYMOLOGY
Ailanthus: from tree of god; *altissima*: very tall

PLACE OF ORIGIN
China

Description   Ailanthus belongs to a mostly tropical plant family. And it looks like a tropical plant: Eleven to 40 dagger-like leaflets adorn each 1- to 3-foot-long compound leaf. Branches grow high on the pale gray trunk. In spring, large clusters of yellow-green flowers appear. From summer into fall, ailanthus stands out thanks to hanging clusters of yellowy to rust-colored, winged samaras. Foliage turns yellow in autumn.

Size   Ailanthus grows to 80 feet or sometimes taller, although heights of 40 to 60 are more common. The trunk can grow up to 2 feet in diameter.

Common Locations   Reproducing both by seeds and by root sprouting, ailanthus is a pioneering, fast-growing tree. In the city, you'll find it in vacant lots, parks, and neighborhoods as well as along fence lines and roadsides.

Notes of Interest   This is the tree mentioned in Betty Smith's famous novel, *A Tree Grows in Brooklyn*.

Ecological Role   Widely introduced from central China starting in the late 1700s, ailanthus now thrives in most of the United States and in many other countries. It is considered an invasive nonnative plant, often growing in large stands that crowd out native vegetation. The tree releases toxins that inhibit the growth of other plants. Some songbirds eat ailanthus seeds and deer sometimes browse young trees. The ailanthus webworm moth is a colorful New World moth that now uses this exotic tree as a host for its caterpillars. Some entomologists think this adaptation enabled the moth to expand its range northward from the subtropics.

KEY POINTS

- Ailanthus is dioecious, with both male and seed-producing female individuals.
- Ailanthus leaf scars—sites from which leaves have fallen—resemble grinning clown faces.
- Snapped branches and flowers of ailanthus emit a strong nutty odor.

Ailanthus

Plate 74

AILANTHUS

American Elm: *Ulmus americana*

ETYMOLOGY
Ulmus: elm; *americana*: of America

PLACE OF ORIGIN
Native

Description   A stately shade tree with sweeping branches that overhang streets and pathways. The American elm reminds many people of small-town America. Some liken this tree's shape to that of a giant bouquet spilling out of a skinny vase. The 3- to 6-inch-long leaves taper at the tip and they have prominent veins and toothed edges, recalling ridged potato chips. American elm's early spring flowers are tiny and green and seldom noticed, as they are high in the tree and because it takes decades before the trees flower and fruit. In late spring, fuzzy-edged fruits droop in small green clusters before falling to the ground.

Size   A large tree, the American elm can grow to more than 100 feet tall, with a trunk diameter of up to 4 feet.

Common Locations   In 2005, 88 young American elms were planted in front of the White House, along Pennsylvania Avenue. More than 8,500 American elms grace the city's streets, and more than 2,500 line parts of the Mall and other National Park Service lands in the city. The Jackson elm, a huge specimen about 200 years old, grows in an alley between Q and Corcoran Streets, N.W., near Dupont Circle.

Notes of Interest   Elm pollen contributes to the impressive pollen counts in Washington, D.C., along with that of maples, oaks, and others.

Ecological Role   One of the most important hardwoods of eastern forest ecosystems, the American elm produces seeds that feed goldfinches and other birds, mice, gray squirrels, and other wildlife. White-tailed deer and cottontail rabbits sometimes browse the twigs and sapling buds.

KEY POINTS

- American elm is the quintessential street tree, planted in towns and cities across the country.
- Starting in the early 1930s, Dutch elm disease was unwittingly introduced into the United States. Over several decades, the beetle-borne fungi killed many of the country's street-side American elms. In the late 1950s, an estimated 38,000

American Elm

Plate 75

AMERICAN ELM

American elms grew in Washington, D.C., far more than remain today.

- Careful crossing of American and disease-resistant Asian elms produced the resilient elms planted today.

American Beech: *Fagus grandifolia*

ETYMOLOGY

Fagus: Latin from Greek referring to edible nuts; *grandifolia*: with large leaves

PLACE OF ORIGIN

Native

Description   American beech is a characteristic tree of the city's deciduous forests. Rarely seen as a street tree, it grows best in undisturbed, moist but well-drained soils. The bark is smooth and light gray and the tree's dense canopy often blocks enough light to prevent other plants from growing beneath the tree. Leaves are spearhead shaped, with toothed edges and sharp-looking tips. Foliage turns yellow in fall. Some dead leaves remain on the branches through winter, especially in younger trees. Growing from slim, slightly zigzagged twigs, the winter buds resemble long, straight, and pointy spines. One to three beechnuts are housed within a prickly husked fruit.

Size   American beech can reach over 80 feet tall with a trunk diameter up to 3 feet.

Common Locations   Rock Creek Park, with its mature forest, is one of the best places to find American beech, but it can also be seen in the mature forests flanking the C&O Canal, at Glover-Archbold Park, Dumbarton Oaks, the U.S. Capitol, and Fort Dupont Park. Valleys and north-facing slopes, which are cooler and wetter, often have larger numbers of this tree.

Notes of Interest   American beech's smooth, gray bark often provides a message board for lovesick youngsters with pocketknives (even though such forms of expression are forbidden in the parks).

Ecological Role   White-tailed deer, gray squirrels, chipmunks, blue jays, wood ducks, and other wildlife eat this tree's oil-rich, triangular beechnuts. Squirrels and far-flying blue jays plant many beechnuts when they store them underground and forget to retrieve them.

KEY POINTS

- Beach Drive, which winds through Rock Creek Park, is named for the river's wide banks rather than this tree.
- The European beech is planted as an ornamental in the area. It is similar but has larger teeth along the leaf edges and fewer pairs of veins on the leaves.

American Beech

Plate 76
AMERICAN BEECH

- American beech twigs often take on a zigzag look. This, along with the pointy buds, gray bark, and spearhead-shaped leaves make it an easy tree to identify.

Sugar Maple: *Acer saccharum*

Description   With blazing scarlet, orange, and yellow fall foliage, the sugar maple is also a main source of maple syrup produced in the Northeast and eastern Canada. Although native to areas just north and west (it is state tree for New York, Vermont, West Virginia, and Wisconsin), it thrives in the nation's capital as a street tree. This is the quintessential maple leaf—five lobes with pointed tips. The paired, winged fruits hang down like inverted U's, or horseshoes. This tree grows to 100 feet tall with a trunk diameter reaching 3 feet.

Ecological Role   Gray squirrels may gnaw at the bark and tap this tree's famous sap. Birds feast on the fruits, which mature in spring. Deer browse buds, twigs, and leaves.

Red Maple: *Acer rubrum*

Description   One of the most familiar forest trees from eastern Canada to Florida, the red maple is also a staple street-side ornamental. It grows to 90 feet tall with a trunk diameter up to 5 feet. From the latter part of October into November, the 2- to 4-inch-long leaves put on quite a show of blazing red and orange. This tree's twigs are reddish as well. The leaves usually have three notched lobes. Young red maples have smooth gray bark; more mature trees have darker, flaky bark. This is the most commonly seen maple in the city, along with the introduced Norway maple. These trees flower starting in late winter, before leaves emerge. At this time, the limbs appear to be covered in a characteristic red haze from the masses of tiny flowers. Red maples contribute to the city's often impressive pollen counts. As leaves develop, the paired red "helicopter" fruits mature, hanging down in a horseshoe-like configuration until they fall in late spring.

Not as common as a street tree, *silver maple* is also native, found in moist, low areas, such the swampy interior of Roosevelt Island. It has deeper-notched leaves, which turn yellow green in fall and long twin fruits that hang in a sharp inverted V.

Ecological Role   Red maple twigs, young leaves, and saplings are favorite forage for white-tailed deer and eastern cottontail rabbits, while gray squirrels eat the buds, fruits, and bark. Some songbirds eat the seeds and buds.

Plate 77
From top: SUGAR MAPLE, RED MAPLE,
NORWAY MAPLE

Norway Maple: *Acer platanoides*

Description   The almost star-shaped leaf of the Norway maple and its smooth edges set it apart from the similar but more jagged leaves of red and sugar maples. The leaves are longer than that of red maple—4 to 7 inches—and have five to seven lobes. The densely foliated, rounded crown of this tree blocks most light below, inhibiting the growth of other plants below it. A widely planted street tree that tolerates a variety of soil and air-quality conditions, Norway maple grows as tall as 100 feet, although such a height is rare, with a trunk diameter that may surpass 2 feet. In late spring, thick, green, winged fruit set in fused pairs point down like shallow, inverted Vs. In fall, the tree's leaves turn yellow or yellow green.

Ecological Role   This tree, originally from Europe, provides shelter and nesting places for city songbirds. It is branded an invasive nonnative plant because its samaras readily sprout after drifting down from the tree's branches, and the resulting trees crowd out native plants.

Sugar Maple

Red Maple

Norway Maple

Sassafras: *Sassafras albidum*

Description   Sassafras is a characteristic small- to medium-sized tree of newly grown forest and forest edge but is also found inside the forest. It reaches 30 to 60 feet tall, with a trunk diameter rarely reaching 3 feet. The leaves are distinctive and often occur in three forms on the same tree—oval, mitten-shaped (two lobes), and trident. The reddish to gray bark on large trees is thick, rough, and ridged. Scraped bark and snapped twigs give off a pleasing, somewhat root beer–like smell. Sassafras branches tend to be wavy, giving a chaotic appearance to the tree crown. Small yellow-green flowers appear March to April, preceding the leaves. Sassafras's fall foliage punctuates the forest with splashes of orange, red, or yellow, and sometimes also purple or pink.

Ecological Role   From late summer to fall, sassafras's dark blue fruits feed songbirds and gray squirrels. Cottontail rabbits eat the bark in winter, and white-tailed deer browse twigs and leaves.

Boxelder: *Acer negundo*

Description   This rangy but interesting small- to medium-sized maple grows in abundance in damp soils, along the city's waterways and in vacant lots and other disturbed areas. Look for it at low points in Rock Creek Park and along the muddy banks of the C&O Canal and Potomac River, among other places. Boxelder's compound leaves have three to five leaflets. They very much resemble poison ivy, which has three leaflets, and ash leaves. This tree rarely lives longer than 100 years and grows up to 60 feet tall, with a trunk up to 2.5 feet in diameter. Compound leaves can be up to 15 inches long and the twigs are a distinctive green to blue green. The gray, deeply ridged bark resembles that of ash trees, but the boxelder has typical maple flowers and paired, winged fruits that hang in clusters.

Ecological Role   White-tailed deer browse twigs and leaves, while the seeds feed mice, squirrels, and songbirds, including grosbeaks.

Sassafras

Boxelder

Plate 78
From top: SASSAFRAS & BOXELDER

Willow Oak: *Quercus phellos*

Description   This large- to medium-sized tree grows in low, usually moist areas from New Jersey to Georgia and west to east Texas. In the city, it grows wild in Coastal Plain sites, such as along the lower Anacostia River. Also widely planted as an ornamental, willow oak lines some streets, parts of the National Zoo's Olmsted Walk, and Roosevelt Island's monument plaza. This tree cannot be confused with other oaks because of its narrow, nonlobed, willow-like leaves. Although tall and upright, this mighty tree has very small acorns. Willow oaks grow to more than 80 feet tall, with huge branches that have fine twigs at the tips. The trunks grow to 3 feet in diameter and are covered with dark-brown bark that is lightly ridged but not flaky.

Ecological Role   Gray squirrels, white-tailed deer, mallards and wood ducks feed on the fallen acorns, eastern kingbirds and other songbirds nest in its branches, and warblers feed on insects in the crowns.

Chestnut Oak: *Quercus montana*

Description   Like the willow oak, the chestnut oak reaches 80 feet tall and a trunk diameter up to 3 feet. Otherwise, this tree is very different: The leaves are shaped like large (up to 8 inch long) jagged teardrops. The acorns are large and oval shaped, and the bark is deeply etched with longitudinal ridges that often form V shapes as they run down the trunk. Unlike the lowland willow oak, the chestnut oak grows on drier slopes and ridgetops. Watch for it in Rock Creek Park and Fort Dupont Park.

Ecological Role   Gray squirrels and white-tailed deer are among the wildlife that eat this tree's acorns.

Willow Oak

Chestnut Oak

Plate 79
From top: WILLOW OAK & CHESTNUT OAK

Eastern White Oak: *Quercus alba*

Description   The characteristic large oak of the region, the eastern white oak grows up to 115 feet tall, with a straight, thick trunk that can reach almost 5 feet in diameter. The bark is thick and shaggy and light gray to whitish, with deeply etched ridges running down the trunk. The leaves grow up to 8.5 inches long and have seven to nine rounded lobes, giving a kind of outward-pointing, rounded fishbone silhouette. A number of large white oaks grace the city. In 2006, Casey Trees, an organization dedicated to conserving Washington, D.C.'s trees, declared the Northampton white oak (in the 2800 block of Northampton Street, N.W.) as the city's largest tree. It is 105 feet tall. Another huge eastern white oak, this one 100 feet tall, grows at Frederick Douglass National Historic Site (Cedar Hill) and is now considered the city's third-largest tree. Herbert Hoover and Franklin D. Roosevelt had white oaks planted at the White House. See also the National Zoo, Dumbarton Oaks and Gardens, Montrose Park, and the U.S. Capitol grounds.

Ecological Role   Raccoons, gray squirrels, and white-tailed deer feed on white oak acorns. In spring, the small flowers attract insects that in turn draw in warblers, flycatchers, and other songbirds.

Northern Red Oak: *Quercus rubra*

Description   A stately tree of the native deciduous forest, the northern red oak grows to 80 feet tall with a trunk up to almost 5 feet in diameter. Smooth gray bark characterizes younger trees but the bark grows rougher and much darker, to almost black, in older trees. Leaves, which are dark green above, paler green below, grow up to 8 inches long and up to 5 inches wide, with sharp-tipped lobes and sinuses (the "valleys" between the lobes) that go about halfway to the rib of the leaf. The acorns are large and round. Northern red oak is a street tree in some parts of the city, including Cleveland Park. One grows on the White House grounds, where Dwight D. Eisenhower had it planted in 1959.

Ecological Role   White-tailed deer, raccoons, gray squirrels, mice, blue jays, and other animals eat the acorns of this oak. Its canopy provides shelter and nesting areas for birds.

Plate 80

From top: EASTERN WHITE OAK, NORTHERN RED OAK, PIN OAK

Pin Oak: *Quercus palustris*

Description   The pointy shoots, "pin"-tipped leaves, and this tree's distinctive shape set the pin oak apart. Branches are set at different angles on different parts of the tree: low branches angle downward, middle branches are horizontal, and higher branches reach upward. A widely planted ornamental and wild native tree in Washington and its suburbs, pin oak grows up to 80 feet tall with a maximum trunk diameter of about 3 feet. Unlike eastern white oak and northern red oak, its leaves grow only up to 5 inches long. They have five to seven lobes separated by wide, U-shaped sinuses. Small, flattish acorns fall below the tree in autumn. Pin oak grows wild from the Mid–Atlantic west to eastern Oklahoma and Kansas. Unlike the eastern white oak, the pin oak rarely lives past 200 years. Look for ornamentals at the U.S. Capitol and White House grounds, around the National Mall, at Hains Point in East Potomac Park, and along residential streets.

Ecological Role   Mallards, wood ducks, blue jays, gray squirrels, and raccoons are among the animals that eat pin oak acorns. Many songbirds nest in the leafy crowns and middle branches of this tree.

Eastern White Oak

Northern Red Oak

Pin Oak

Ginkgo: *Ginkgo biloba*

ETYMOLOGY
Ginkgo: of Japanese from Chinese meaning silver apricot; *biloba*: having two lobes

PLACE OF ORIGIN
China

Description   This strange tree, also called the maidenhair tree, is a widely planted ornamental that is the sole surviving species of a unique plant line dating back to the Permian period, about 300 million years ago. The fan-shaped leaves have obvious parallel veins. Male trees are upright and produce small, abundant cones. Female trees tend to have wider crowns and more scattered cones that hang from long stalks. In autumn, ginkgo leaves turn yellow and drop. The bark on older trees is strongly ridged and rough. Even after its leaves drop, a bare ginkgo tree is distinctive, with finger-like twig spurs that jut off the branches and a twisted, candelabra-like arrangement of branches off the main trunk.

Size   Ginkgo trees grow up to 100 feet tall. They are slow growing.

Common Locations   A large ginkgo grows behind the Smithsonian Castle and another grows in front of the U.S. Capitol. Gingko is also planted along streets in Chinatown and Adams Morgan.

Notes of Interest   In Asia, gingko has long been used as an herbal remedy. Among other things, it likely improves circulation and is believed by some to improve memory.

Ecological Role   A tree that resists pests and diseases and thrives where many trees cannot—in polluted, compacted, and otherwise compromised city soils. Some songbirds nest in its branches, and it provides much-needed shade in the oppressive Washington summers.

KEY POINTS

- In fall, female trees produce fleshy yellow or orange fruits that when ripe or overripe give off a nasty odor. For this reason, male trees are often planted along streets and walkways.
- In China and Japan and in Asian markets in North America, ginkgo seeds are roasted, sold, and eaten.
- Ginkgo is the world's oldest tree species according to *Guinness World Records*.
- It has been cultivated in Asia for many centuries; no one is sure if any purely wild specimens exist. The first U.S. specimen was likely one planted in Philadelphia in 1784.

Ginkgo

Plate 81

GINKGO

Crape-Myrtle: *Lagerstroemia indica*

ETYMOLOGY

Lagerstroemia: named for Swedish merchant Magnus von Lager-stroem, who supplied first specimens of this plant to Carl Linnaeus, Swedish naturalist and the father of animal and plant classification; *indica*: from India

PLACE OF ORIGIN

Asia, probably China

Description   Crape-myrtle is considered a southern tree by many. It cannot sustain prolonged, bitterly cold winters, although it is planted in coastal areas as far north as Cape Cod. It also flourishes along much of the West Coast. This small tree or large shrub is very popular in the city due to its yard-friendly size and profusion of flowers from July into September. Crape-myrtle grows fast and usually branches into several main trunks unless pruned to form a more typical tree shape. The deciduous leaves are simple in shape and short—only up to 3 inches long. Flowers have six petals each and bloom in thick clusters at branch tips. There are various shades of red, pink, and purple as well as white. The bark is grayish brown, with strips peeling to reveal orange to greenish tones. Clusters of brownish fruits appear in late summer and fall. Autumn foliage color may be orange, yellow, or red.

Size   Crape-myrtle rarely grows taller than 30 feet, with a canopy spread up to 25 feet.

Common Locations   A wide variety of crape-myrtle cultivars are planted, labeled, and growing at the National Arboretum's Gotelli Collection of Dwarf and Slow Growing Conifers and near the administration building and visitor center. You will also see them growing in many front yards.

Notes of Interest   Crape-myrtle belongs to the loosestrife family.

Ecological Role   Some bees and butterflies come for the flow-ers' nectar and American robins, northern mockingbirds, and northern cardinals sometimes place their nests in crape-myrtle branches.

KEY POINTS

- While not a common ornamental much farther north than Baltimore and the coast, hardiness zones have moved north in recent years, and crape-myrtle may soon become a more fam-iliar sight in northern areas where they were not grown before,

Crape-Myrtle

Plate 82
CRAPE-MYRTLE

particularly in cities, where pavement, buildings, and other development create a warm microclimate.
- Crape-myrtle grows in a wide variety of soil conditions—those with clay, loam, and sand, and both in acidic and alkaline conditions.
- The National Arboretum has developed a wide variety of disease-resistant cultivars of crape-myrtle.

Mulberry: *Morus* spp.

ETYMOLOGY
Morus: mulberry

White Mulberry Tree: *Morus alba*

PLACE OF ORIGIN
Eastern Asia

Red Mulberry Tree: *Morus rubra*

PLACE OF ORIGIN
Native

Description A mulberry tree may have some leaves with no lobes and others with two, three, or five lobes. Leaves are toothed around the edges and sharp tipped, and grow up to 7 inches long. Red mulberry leaves are usually rough on the upper surface, while on white mulberry, this surface is smooth and glossy. The oblong fruits in red mulberry are purple when ripe, but ripe fruit colors vary in white mulberry, including white, pink, purple, and black. The red mulberry tends to have a more open crown with wispy branches; white mulberry usually has a more closed crown with dense clusters of twigs. These species often hybridize, and hybrids have features intermediate between the two.

Size Although often considered small trees, mulberries can grow to more than 60 feet tall, with trunks up to almost 5 feet in diameter.

Common Locations Red mulberry is a common native tree in eastern lowlands as well as on damp hillsides and in fields. A native of China, the white mulberry is considered an invasive nonnative plant. It is now found in most states and tends to be more common than the red in developed areas. Some large white mulberry trees grow near the Washington Monument. Both species occur in the city.

Notes of Interest Mulberries, with their varied leaf shapes, may be confused with sassafras, which may also have no lobes, two lobes, or three. But mulberry leaf edges are toothed.

Ecological Role A fruiting mulberry tree rarely holds its crop for long. In early summer, catbirds, mockingbirds, American robins, cedar waxwings, downy woodpeckers, Baltimore orioles, and many other birds relish the ripe berries, as do people, raccoons, gray squirrels, and white-tailed deer.

Mulberry

Plate 83

MULBERRY

KEY POINTS

- White mulberry was introduced during efforts to establish a silkworm industry in the United States during colonial times.
- White mulberry frequently hybridizes with red mulberry.

Eastern Redbud: *Cercis canadensis*

ETYMOLOGY
Cercis: Greek for weaver's shuttle, referring to shape of seedpods; *canadensis*: of Canada

PLACE OF ORIGIN
Native

Description The wide, heart-shaped leaves of the redbud are distinctive, as are the haze of light to dark pink, tiny flowers that appear before the leaves in early spring. The smooth-edged leaves grow as long as 5 inches and have short tips. A member of the legume family, redbud has flat 2- to 3-inch-long seedpods, which develop in summer and drop to the ground by autumn along with the tree's yellowed leaves. The bark on older trees can be scaly or smooth and is gray, infused with reddish orange. This tree usually has several main trunks that branch off from close to the ground. Unlike the trunk, twigs are darker and smooth. This tree's crown is messy, with a tangled arrangement to the branches.

Size Redbud is a small tree, rarely reaching a height of more than 25 feet, with a trunk up to 12 inches in diameter.

Common Locations This tree grows wild and as a landscaping accent in many places, including at the National Arboretum, the National Zoo, and Rock Creek Park.

Notes of Interest Eastern redbud has close relatives in California, the Mediterranean, and China. The Chinese species also is planted as an ornamental in Washington, D.C. Despite its species name *canadensis*, eastern redbud grows naturally in just a few areas in Canada near the U.S. border.

Ecological Role Redbud, along with flowering dogwood and spicebush, often makes up an important part of the forest understory. Some birds nest in its branches, bees find nectar in its blooms, and gray squirrels, deer, cardinals, and other creatures sometimes eat the seeds.

KEY POINTS
- Washington, D.C.'s redbuds bloom in April, before most deciduous trees are in full leaf.
- The flowers appear in clusters along small branches.
- Redbuds often don't live longer than 50 years.

Eastern Redbud

Plate 84

EASTERN REDBUD

Sweetgum: *Liquidambar styraciflua*

ETYMOLOGY
Liquidambar: liquid amber, referring to its yellow sap; *styraciflua*: aromatic resin

PLACE OF ORIGIN
Native

Description   This common tree is known for its five-pointed, star-shaped leaves and the spiky fruit balls that hang from its branches. Some of these balls remain on the tree through winter, well after all the leaves have fallen. Sweetgum foliage puts on a fall show, varying from yellow to red purple. The thick gray bark is deeply etched with furrows running up and down the trunk. Main branches jut directly out from the very straight trunk at almost right angles.

Size   Sweetgum is a large tree, growing to 80 feet or taller and with a trunk diameter up to 4 feet.

Common Locations   Sweetgum is found in the city's forests but is also planted as an ornamental or allowed to grow as a yard tree. Watch for it along the River Trail at Kenilworth Aquatic Gardens, on the U.S. Capitol grounds, and in West Potomac Park, including Constitution Gardens.

Notes of Interest   In woodlands, sweetgum is often found growing alongside northern red oak, tuliptree, willow oak, pin oak, and other trees.

Ecological Role   Chickadees, cardinals, finches, and goldfinches are among the birds that feed on the seeds found in sweetgum balls. Gray squirrels and eastern chipmunks also eat the seeds.

KEY POINTS
- Sweetgum is a staple tree of the southern lumber industry.
- A gum made from this tree's sap and inner bark was once chewed and also used to give aroma to soap and other products. Native Americans made medicinal teas by boiling the bark.

Sweetgum

Plate 85

SWEETGUM

American Sycamore: *Platanus occidentalis*

ETYMOLOGY
Platanus: plane tree; *occidentalis*: western

PLACE OF ORIGIN

Native

Description   The huge sycamore catches the eye because of its sheer size and its exposed white inner bark decorated with strips of peeling gray and cinnamon outer bark. The leaves are large—up to 8 inches long—and shaped somewhat like maple leaves but with shallower lobes. The colossal branches reach left and right like massive arms. The fruit clusters are knobby balls, not spiky as in the sweetgum.

Size   One of the largest eastern trees and one of the widest in the country, the American sycamore can live for centuries and reach up to 175 feet tall, with a trunk diameter up to 8 feet.

Common Locations   American sycamores grow along shores of the Potomac at such places as the C&O Canal and Roosevelt Island. They are also found along Rock Creek, at Fort Dupont Park, and at Glover-Archbold Park.

Notes of Interest   There are also sycamore species native to streams and gullies in California and Arizona.

Ecological Role   American sycamores often grow along the edges of waterways and provide important nesting sites for wood ducks, woodpeckers, and many songbirds, including orioles and warblers.

KEY POINTS

- The similar London plane-tree, a cultivated hybrid, is also planted in some places in the city as a pollution-tolerant shade tree.
- American sycamore grows in moist soils across most of the eastern United States.

American Sycamore

Plate 86

AMERICAN SYCAMORE

Tuliptree: *Liriodendron tulipifera*

ETYMOLOGY
Liriodendron: Greek for lily tree; *tulipifera*: Latin for tulip bearing
PLACE OF ORIGIN
Native

Description The tuliptree holds beautiful surprises at all seasons. The trunk is straight and round with tight-looking bark in two tones of gray. The crown is rounded but grows expansive in older trees. In spring, the crown is adorned with large, showy pale-yellow flowers, each seemingly painted with a wide orange band. In summer, the green, upward-pointing, rocket-shaped seed cones can be seen. These eventually dry out, and from late fall into winter, the winged seeds or samaras within separate and fall. The somewhat hat-shaped, four-lobed leaf is distinctive and is as wide as it is long.

Size A large tree, the tuliptree reaches heights of 120 feet with a trunk diameter of 6 feet.

Common Locations The city's second-largest tree is a 96-foot-tall tuliptree growing in Montrose Park, north of Q Street, N.W., in Georgetown. Tuliptree grows in most D.C. wooded parks and is found on the U.S. Capitol grounds.

Notes of Interest The tuliptree is one of the first native species to begin leafing out in spring. It is also one of the first tall-growing hardwoods to colonize cleared areas, its shade providing shelter for less sun-tolerant plants.

Ecological Role In spring, the broad, tulip-like flowers attract ruby-throated hummingbirds high in the tree's crown. Bees feed on the flowers' nectar as well, and many birds and small mammals eat the seeds, including gray squirrels, finches, and cardinals. The leaves feed tiger swallowtail caterpillars. White-tailed deer feed on the saplings and twigs. Scarlet tanagers and migrating rose-breasted grosbeaks are among the songbirds that find shelter and food among the high branches.

KEY POINTS

- The state tree of Indiana, Kentucky, and Tennessee.
- One of the city's tallest trees.
- Although often called yellow poplar and tulip poplar, this tree belongs to the magnolia family.

Tuliptree

Plate 87

TULIPTREE

Southern Magnolia: *Magnolia grandiflora*

ETYMOLOGY

Magnolia: named for Pierre Magnol, a French botanist; *grandiflora*: with large flowers

PLACE OF ORIGIN

Southeastern United States

Description   An evergreen with an oval shape, this medium-sized tree is native from the Dismal Swamp region at the Virginia–North Carolina border across and down to Texas. But its attractive appearance and year-round greenery make it a popular ornamental in Washington, D.C. The elliptical leaves grow to 8 inches long and are dark green and glossy above and pale green below, often with rusty hairs. Large, showy, sweet-smelling white flowers decorate this tree from late spring into summer, followed in fall by reddish, hairy seed cones that grow up to 10 inches long. During winter, seed cones fall and collect beneath the tree.

Size   Southern magnolia grows up to 90 feet tall, and its trunk can reach 4 feet in diameter.

Common Locations   In the city, southern magnolia adorns the outside of many office buildings and famous buildings, including the White House, the Lincoln Memorial, the U.S. Capitol, the Smithsonian Castle, Hirshhorn Museum on the National Mall, and Frederick Douglass National Historic Site (Cedar Hill).

Notes of Interest   The oldest commemorative plantings growing at the White House are two southern magnolias dating back to 1830. They grow on the south-facing left side of the White House. Many believe Andrew Jackson brought these trees from his Tennessee home, the Hermitage, in memory of his wife, who died shortly before he took office. Until a revision in 1998, 20-dollar bills featured the White House's south side, showing the magnolias blocking much of the view of the building's left side. Today, the building's north side is depicted.

Ecological Role   Gray squirrels, opossums, mice, chipmunks, and seed-eating birds sometimes eat this tree's seeds. In its native forests, southern magnolia grows in association with species that also commonly grow in Washington, D.C., including sweetgum, tuliptree, and American beech.

TREES

Southern Magnolia

Plate 88
SOUTHERN MAGNOLIA

KEY POINTS
- In their native range, southern magnolias often grow in shade. This attribute, along with its showy leaves and flowers, makes it a natural choice for planting between D.C.'s office buildings and museums.
- Southern magnolia has also been planted in cities around the world.

Weeping Willow: *Salix babylonica*

ETYMOLOGY
Salix: willow; *babylonica*: of Babylon
PLACE OF ORIGIN
China

Description   An elegant tree with a broad crown and "weeping," drooping branches. Finely toothed, slender leaves are bright green and up to 7 inches long, but only about a half-inch or less wide. The drooping twigs and leaves conceal from view much of the tree's rough, gray bark. In spring, inch-long catkins appear on the branches.

Size   Weeping willows grow up to 60 feet tall, and the trunk can grow up to 7 feet in diameter.

Common Locations   Anacostia Park, East and West Potomac Parks, and the National Arboretum are among the places you will see this tree.

Notes of Interest   Carl Linnaeus, who started the classification system used today for wildlife, named this tree for the willows of Babylon mentioned in the Bible. Botanists now believe, however, that the biblical trees were poplars.

Ecological Role   This widely planted ornamental provides shade in open areas and its roots absorb rainwater and protect soil from washing away.

KEY POINTS
- Like the native black willow, weeping willows favor damp locations.

Weeping Willow

Plate 89
WEEPING WILLOW

Black Cherry: *Prunus serotina*

ETYMOLOGY

Prunus: Latin, from Greek for plum tree; *serotina*: late flowering or late ripening

PLACE OF ORIGIN

Native

Description   Black cherry's spearhead-shaped leaves and the long, somewhat grape-like clusters of small white, five-petaled flowers catch the eye in spring. The flowers are followed by red fruits that ripen to black by summer. Branches are blackish and smooth, with telltale white lines, while the bark on the trunk of older trees is dark and flaky. The leaves are fringed with rounded teeth. Autumn leaf colors vary from yellow to orange to a dull scarlet.

Size   Black cherry grows up to 100 feet tall with a trunk diameter sometimes reaching four and a half feet. A height of around 60 feet tall is about average.

Common Locations   Black cherry is a common tree in the city's forests and forest edges. Its seeds are distributed in birds' droppings so saplings often pop up in meadows, parks, and backyards, where the trees are sometimes left to grow.

Notes of Interest   Washington has many types of nonnative flowering cherry trees planted as ornamentals. Sweet or mazzard cherry, from Eurasia, now also grows wild. It flowers earlier than black cherry, as early as March. The fruits are larger, sweet, and don't hang in branched clusters like black cherry. The leaves are also broader and the flowers much larger.

Ecological Role   Fallen black cherries form part of the red and gray foxes' summer diet. The pits can often be seen in the animals' scat. Opossum, raccoon, eastern chipmunk, and mice also eat black cherries, as do many birds, including cedar waxwing, American robin, eastern bluebird, northern flicker, northern mockingbird, gray catbird, blue jay, Baltimore oriole, and American goldfinch. Black cherry is a host tree for the caterpillars of spring azure, red-spotted purple, viceroy, and tiger swallowtail butterflies, and a favored location for eastern tent caterpillar nests, which attract caterpillar-eaters such as yellow-billed cuckoo and Baltimore oriole.

KEY POINTS

- Black cherry is found throughout the East but also occurs in scattered areas from Arizona and New Mexico south to Guatemala.

Black Cherry

Plate 90
BLACK CHERRY

- In early spring, many cherries have gauzy eastern tent caterpillar nests in their branches, soon followed by thinning leaves, as the caterpillars grow and spread through the trees on their daily forays from the nest. The trees usually regain their foliage quickly after the caterpillars disperse to spin their cocoons elsewhere.

Yoshino Cherry: *Prunus* x *yedoensis*

ETYMOLOGY
Prunus: Latin, from Greek for plum tree; *yedoensis*: of Tokyo

PLACE OF ORIGIN
Japan

Description   For more than a century now, the city's famous flowering trees have drawn tourists to see and smell their prolific white, almond-scented flower clusters. Younger trees have straight trunks with smooth bark and horizontal partial bands, called lenticels. Old trees are stout and knobby.

For yearly bloom predictions and for information on events relating to the annual National Cherry Blossom Festival, see the National Park Service, National Cherry Blossom Festival, or *Washington Post* websites. Blooming Yoshino trees can be seen for up to two weeks, but the peak bloom—when about 70 percent of trees are flowering—lasts just a few days. According to the National Park Service, the average peak bloom date is April 4. The earliest recorded peak date was March 15, 1990. In 2012, following a very warm winter, trees began blooming mid-March and peaked around March 20. The latest peak bloom date was April 18, 1958. Once trees reach their peak, wind can quickly carry away clouds of blooms, or driving rain may wash them off the trees, especially if there is frost or unusually warm temperatures.

Size   Yoshino cherry trees rarely reach 50 feet tall; 30 feet is a more typical height.

Common Locations   The Tidal Basin (West Potomac Park) is the city's premiere area for Yoshino cherries, where they were originally planted and where the largest assemblage remains. Virtually all of the Tidal Basin's cherry trees are Yoshino cherries. They are also found in small groves around the Washington Monument and at Dumbarton Oaks and Gardens' Cherry Hill. Some grow in East Potomac Park, but this park is best known as for its many Kwanzan cherry trees, which bloom about two weeks later than the Yoshinos.

Notes of Interest   A gift from Japan to the United States in 1912, 3,020 cherry trees of 12 varieties were planted in the city. Of these, 1,800 were Yoshino. Today, a few gnarled originals remain. The oldest trees sit on the northwest corner of the Tidal Basin, just east of the Martin Luther King, Jr. Memorial. The Jefferson Memorial sits directly across from them, on the far southeast bank of the basin. The National Arboretum works with the National Park Service to

Plate 91

YOSHINO CHERRY

plant new trees propagated from some of the original trees sent from Japan in 1912.

Ecological Role   These celebrity trees represent the goodwill between Japan and the United States and evoke a love of nature in many who visit the city. But they face urban and natural threats. Visitors are asked not to climb the trees, pluck blossoms, or walk around the base of the trees because soil compaction damages their roots. Local wildlife sometimes give the trees unwanted attention. Yoshino cherries host eastern tent caterpillars, aphids, borers, and other insects, and sometimes attract the attention of beavers. Horticulturists carefully monitor and treat the trees to keep them as healthy as possible. Birds eat the small fruits produced by these trees.

KEY POINTS

- The Yoshino cherry is a hybrid developed in Tokyo in the late 1800s.
- Over the years, Japanese and U.S. horticulturists have exchanged stock from different Yoshino cherries to keep the genetic stock strong.
- Just east of the Jefferson Memorial, along the shores of the Tidal Basin, is the Indicator Tree. The National Park Service sets its peak blooming prediction from this tree, which for some reason blooms about a week before the others.

Yoshino Cherry

Flowering Dogwood: *Cornus florida*

ETYMOLOGY
Cornus: refers to hard wood; *florida*: flowering

PLACE OF ORIGIN
Native

Description   The flowering dogwood is one of the region's prettiest native trees. Between April and early May, blooming dogwoods create a white blanket beneath the canopy of taller forest trees. But what look like large white spring flowers are actually petal-like bracts surrounding the cluster of the actual tiny, yellow-green flowers in the center. The dogwood's leaves are distinctive: rounded, with wavy parallel veins and a diagonally pointing sharp tip. In fall, clusters of missile-shaped berries ripen to red, and foliage turns to red, orange, or yellow. In winter, small button-like flower buds decorate the bare branch tips. The brown bark appears broken up into small sections surrounded by deep fissures, giving it an almost latticework appearance.

Size   Flowering dogwood usually grows up to about 30 feet tall but rarely reaches 40 feet tall. The trunk can reach up to a foot in diameter.

Common Locations   Flowering dogwood is found in most of the city's forested parks and neighborhoods. It grows among the azaleas at the National Arboretum's Mount Hamilton, at Lady Bird Johnson Park, and on the White House grounds.

Notes of Interest   To see many other types of dogwoods, check out the dogwood collection at the National Arboretum.

Ecological Role   Many birds nest in flowering dogwoods. Gray squirrels and cottontail rabbits eat the fall fruits, as do migrating wood thrushes, brown thrashers, gray catbirds, cedar waxwings, and robins. The flowers attract butterflies.

KEY POINTS

- Until recently, the flowering dogwood was an abundant forest tree. Numbers declined following the spread of a fungus called dogwood anthracnose.
- Kousa dogwood, from Asia, is also planted as an ornamental, as are hybrids between Kousa and flowering dogwood. Most of these trees resist the fungus.
- In 1915, the U.S. government sent flowering dogwoods to Japan to reciprocate for that country's gift of cherry trees in 1912.

Flowering Dogwood

Plate 92
FLOWERING DOGWOOD

- Flowering dogwood is the state tree of Missouri, the state flower of North Carolina, and both state tree and flower of Virginia.

American Holly: *Ilex opaca*

ETYMOLOGY
Ilex: refers to Europe's holly oak, *Quercus ilex*; *opaca*: shaded, dark
PLACE OF ORIGIN
Native

Description   A familiar tree with a pyramid-shaped crown, spiny evergreen leaves, and red berries in late fall and early winter. Tiny greenish flowers appear in May. The spine-edged leaves are glossy and dark green above and pale yellowish or green below and rarely grow longer than 3 inches.

Size   American holly reaches 50 feet tall with a trunk rarely to 2 feet in diameter.

Common Locations   American holly grows wild within the city's forests, including along the River Trail at Kenilworth Aquatic Gardens, at Roosevelt Island, and at Glover-Archbold Park. It is widely planted as an ornamental throughout the city. Large individuals, for example, ring the Smithsonian Castle and grow by the Lincoln Memorial, Jefferson Memorial and Tidal Basin (West Potomac Park), and the U.S. Capitol.

Notes of Interest   A wide variety of showy hollies can be seen at the National Arboretum's 10-acre Holly and Magnolia Collection.

Ecological Role   About 40 bird species have been documented eating American holly berries. Frequent visitors include flocks of American robins and cedar waxwings, sometimes joined by one or two hermit thrushes. Northern mockingbirds may battle these birds for access to the berries. Through the winter months, American holly's evergreen leaves and boughs shelter many birds and small mammals from snow, ice, and wind. Mourning doves, blue jays, and others place their nests among its thorny leafed branches.

KEY POINTS

- American holly grows wild from coastal Massachusetts south and west to central Texas.
- American holly boughs are often used as Christmas decorations.
- American holly is the state tree of Delaware.

American Holly

Plate 93

AMERICAN HOLLY

NATIVE SHRUBS

Spicebush: *Lindera benzoin*

Description   An abundant shrub that grows in rich soil, often in moist forests bordering rivers and streams. The leaves are simple, blade shaped, and alternate. Crushed leaves, twigs, and fruits exude a sweet aroma. Spicebush is one of the region's earliest flowering shrubs. Sometimes blooming as early as mid-March, the small yellow flowers appear before the shrub's leaves emerge. Fruits ripen in late summer and early fall. Rock Creek Park and Glover-Archbold Park are among many sites where this shrub grows in profusion.

Ecological Role   Spicebush swallowtail caterpillars use this plant as a host. Migrating thrushes and other birds eat the fruits, and eastern cottontail rabbits sometimes nibble the buds, leaves, and twigs.

Mapleleaf Viburnum: *Viburnum acerifolium*

Description   The lobed leaves strongly resemble those of red maple, but this is a low shrub that grows to about 6 feet tall. Also, unlike red maple, the mapleleaf viburnum produces small white flowers from April into June, followed later by reddish purple berries.

Ecological Role   Small mammals and birds eat the berries, and this shrub forms low thickets within the forest shade, providing cover and nesting sites.

Arrowwood Viburnum: *Viburnum dentatum*

Description   The ramrod-straight stems of this understory shrub were used by Native Americans for arrow shafts. The round leaves are toothed and ridged with veins that make them somewhat resemble green, ridged potato chips. Flat-topped white flower clusters show from mid-May to early summer. Blue berries adorn the shrubs by late summer.

Ecological Role   Arrowwood provides cover for nesting birds. White-tailed deer browse the twigs, and birds, including gray catbirds and American robins, eat the berries. Squirrels and rabbits eat both the fruits and bark. This is a host plant for spring azure butterfly caterpillars and a nectar source for adults. The blooms also attract bees.

Spicebush

Arrowwood Viburnum

Mapleleaf Viburnum

Plate 94

From top: SPICEBUSH, MAPLELEAF VIBURNUM,
ARROWWOOD VIBURNUM

NONNATIVE SHRUBS

Bush Honeysuckle: *Lonicera* spp.

Description   These fast-growing, up-to-15-feet-tall Eurasian shrubs dominate the undergrowth in parts of Rock Creek Park and other areas. The oval, short-stemmed leaves run opposite along the branches. Bush honeysuckles flower in spring. Depending on the species or variety, the paired, inch-long, tubular blooms may be white or pink or red. The red or orange fruits mature from summer to fall.

Ecological Role   These shrubs displace native shrubs, but they do provide nest sites for birds. The dense shade they produce may shade out other native plants. Birds and small mammals eat the fruits.

Wineberry: *Rubus phoenicolasius*

Description   A red-stemmed member of the raspberry genus, now more commonly seen in the city than either native blackberry or raspberry. The reddish coloration is actually provided by abundant red "hairs." Leaves are composed of three purple-veined, toothed leaflets. In summer, songbirds eat the bright-red fruits, which humans also find tasty. Wineberry grows in dense clusters in many of the city's woods.

Ecological Role   This bush crowds out native vegetation but provides abundant food for birds such as mockingbirds, catbirds, and small mammals, which in turn help spread it to new areas via their droppings. The bush also spreads via its root buds and when branches touch the ground and take root.

Multiflora Rose: *Rosa multiflora*

Description   A thorny, small-flowered Asian shrub, multiflora rose grows in a variety of habitats. It produces abundant fruit, or rose hips, from late summer into winter. The white to pink inch-wide flowers grow in fragrant clusters.

Ecological Role   The fruits provide food for birds such as the northern mockingbird. This shrub's dense, thorny tangles shelter bird nests and small mammals including eastern cottontail rabbits. The flowers attract bees and other pollinating insects.

Bush Honeysuckle

Wineberry

Multiflora Rose

Plate 95
From top: BUSH HONEYSUCKLE,
WINEBERRY,
MULTIFLORA ROSE

NATIVE VINES

Poison Ivy: *Rhus radicans or Toxicodendron radicans*

Description   A member of the cashew family that strikes fear in the hearts of hikers. Most people get a rash after contacting any part of this plant, be it the glossy, alternative leaves consisting of three spearhead-shaped leaflets, the smooth stems, the hairy vines, or the spidery roots. Leaves turn red, orange, or yellow in autumn. Often the leaves have smooth edges; sometimes they have some notches. Leaves resemble those of the boxelder, although this tree has three to five leaflets per leaf. Virginia creeper is also similar but always has five leaflets per leaf.

Ecological Role   Many birds, including the yellow-rumped warbler and Carolina chickadee, eat the small greenish or whitish poison ivy berries and forage among the hairy vines and stems to find active (in warm months) and dormant (in cold months) invertebrates.

Virginia Creeper: *Parthenocissus quinquefolia*

Description   A thin but ropey member of the grape family that climbs tree trunks, fences, and the sides of houses. Although very much like poison ivy in its growing habit, this vine is not hairy and always has five, not three, leaflets per leaf. In fall, Virginia creeper foliage turns scarlet and burgundy. Ripe fruits are blackish. Virginia creeper is more or less as common as poison ivy and found in similar locations—at the forest edge, along fence lines, or in open or regenerating forest.

Ecological Role   Virginia creeper fruits provide important fall and winter food for birds and small mammals. Wrens, thrashers, and other shy songbirds skulk or nest within its foliage.

Virginia Creeper

Poison Ivy

Plate 96

From top: POISON IVY & VIRGINIA CREEPER

NONNATIVE VINES

Porcelainberry: *Ampelopsis brevipedunculata*

Description   In the city, this wild grape look-alike thrives in moist soils that get adequate sunlight. A member of the grape family, it climbs 20 feet or higher in trees or up telephone poles. Leaves on the same plant may be heart shaped or strongly lobed. Many find it hard to differentiate between this plant and native wild grape but when snapped apart you can see that the pith of porcelainberry is white while that of wild grape is brown. In fall, berries are easy to identify: Unlike wild grape, which has dark fruits, porcelainberry fruits vary in color and may be white, yellow, green, pink, purple, or light blue.

Ecological Role   The vine creates dense tangles and is a fast grower, smothering nearby shrubs and small trees. Native wild grape is not nearly as vigorous. Birds and small mammals eat the fruits.

Japanese Honeysuckle: *Lonicera japonica*

Description   This Asiatic vine belongs to the same genus as the bush honeysuckles. The thin, somewhat wavy white flowers are fragrant, especially by late spring and early summer. Aged flowers turn a pale yellow. The leaves are oval and dark green and many remain on the vine through winter. The reddish (and later brown and woody) vines wrap tightly around shrubs, young trees, and fence posts. Children often pluck the flowers and pull the long stamens out through the bottom, collecting a single drop of sweet nectar. In late summer and fall, shiny black berries appear.

Ecological Role   This vine's black berries provide some late-summer and fall food for birds, while its flowers attract ruby-throated hummingbirds, sphinx moths, and swallowtails. White-tailed deer and eastern cottontails eat the foliage, especially in winter.

Porcelainberry

Japanese Honeysuckle

Plate 97
From top: PORCELAINBERRY
& JAPANESE HONEYSUCKLE

CHAPTER 6

mushrooms

Artist's Conk: *Ganoderma applanatum*

ETYMOLOGY
Ganoderma: shiny skin; *applanatum:* flat

PLACE OF ORIGIN
North America

Description   The cap of the artist's conk is hard and semicircular or fan-shaped, growing from 2 inches to 2 feet wide. The rim is whitish and the bands may be pale or dark, rust-colored, tan, grayish, or whitish. The underside is bright white.

Common Locations   Forests with large deciduous trees often provide habitat for this mushroom. You may find it on logs, stumps, or dead portions of living trees.

Notes of Interest   Artist's conk is white underneath, but any disturbance to this area, including scraping with a stick, stains touched areas brown.

Ecological Role   This fungus helps decompose logs and stumps, but when it grows on living trees, it can weaken them, paving the way for wood-boring insects, which in turn attract woodpeckers. Artist's conk can last decades. It is one of the longest-living fungi.

KEY POINTS

• Artists use the hard, white underside of this shelf fungus as a palette for hand carving or burning in artwork or for drawing.

• Artist's conk is one of the longest-living fungi.

• While most mushrooms can be easily collected with a knife, an axe is recommended for this tough fungus.

Plate 98

ARTIST'S CONK

Chicken Mushroom, Sulphur Shelf, or Chicken-of-the-Woods: *Laetiporus sulphureus*

ETYMOLOGY

Laetiporus: with bright pores; *sulphureus:* yellow color of sulphur

PLACE OF ORIGIN

Chicken mushroom can be found in many parts of North America and Europe.

Description   This common and colorful mushroom is a feast for the eyes and easy to identify. Each cap grows up to a foot or, rarely, 2 feet wide. The caps overlap, looking somewhat like a fluffy pile of orange-yellow pancakes. Below, the pores are bright yellow. They are called shelf or bracket fungi because they grow flat, with little or no stem, protruding from the wood like brackets or shelves. They can grow very large: Individual shelves may weigh a pound or more each.

Common Locations   From June to November, you may encounter chicken mushrooms growing anywhere in the city where there are trees, logs, or stumps.

Notes of Interest   It should not be eaten raw, but when this easily identified mushroom is properly prepared, it makes a tasty substitute for chicken breast meat, or it can be included in chicken dishes. Some people even think it tastes like lobster. Although experts consider it a relished, safe-to-eat mushroom, some people feel sick after eating it.

Ecological Role   White-tailed deer eat chicken mushrooms, and a wide variety of invertebrates feed on them as well, including flatworms, pillbugs, and specialized gnats, flies, and beetles. The fungus helps breaks down wood, enriches the soil, and attracts these invertebrates, thus further aiding in the decomposition process.

KEY POINTS

- Chicken mushrooms are polypores. They have minute pores beneath their caps, while many other mushrooms have slitted gills.
- This mushroom is best cooked when mature, but before it gets tough and woody.
- Chicken mushrooms can be found growing in the same spot for a number of years.
- A similar mushroom, the white chicken mushroom, grows in spring and summer. It is also edible. It has a pinkish orange cap and is white, not yellow below.

Plate 99

CHICKEN MUSHROOM, SULPHUR SHELF,
OR CHICKEN-OF-THE-WOODS

Turkey Tail: *Trametes versicolor*

ETYMOLOGY

Trametes: thin or skinny; *versicolor:* multicolored

PLACE OF ORIGIN

Turkey tail is found in many temperate parts of the world, including North America and Eurasia.

Description   This mushroom's common name says it all: The shape and colorful bands recall a turkey's tail. The tough, shelf-like caps grow to only 1 to 4 inches wide in overlapped clusters. The bands vary in color and may be gray, white, brown, green, blue, red, or yellow.

Common Locations   Look for this shelf fungus in shady forest in well-wooded parks, where it grows on trees, particularly dead or dying ones. This mushroom may be seen year round.

Notes of Interest   Dried turkey tail can be used to make jewelry and other crafts.

Ecological Role   Turkey tail flourishes on dead wood of deciduous trees. But it can also grow on living trees and can be detrimental to fruit and other trees. It helps decompose wood and enrich the forest soil.

KEY POINTS

- Turkey tail has been used in traditional Asian medicine remedies for centuries and is one of the first mushrooms from which a cancer-treating drug was derived.
- Although it is seen year round, this mushroom grows from May to December. It can live for years.
- Although too tough to eat, turkey tail can be boiled to make a healthful tea or soup.

Plate 100
TURKEY TAIL

7

geology

Sykesville Formation

ETYMOLOGY
Named for a town in Maryland where rocks of this formation have been mined.

Natural History   The Taconian Orogeny occurred around 460 million years ago. This intense mountain building event took place on the East Coast, during the late Ordovician period, when what we would now call North America was a much smaller landmass. The eastern coastline sat far west of what is now the city on a tectonic plate with its edge far to the east, submerged by the ocean. This ancestral "North American" plate slammed into a chain of volcanic islands, and its edge was shoved downward beneath the colliding plate. Similar tectonic action, called subduction, now occurs in Japan, Indonesia, the Philippines, and other areas. As the plate was forced downward, sediments deposited on top of the plate's edge were sheared off and piled up at the contact zone, creating what geologists call an accretionary wedge. The incredible scraping and squeezing of oceanic sediments transformed sediments into metamorphic rock.

Description   Much of the rock you see in the Sykesville Formation is a bedrock type called metagraywacke. This metamorphic rock formed from sandstone (graywacke) scraped and squeezed under elevated heat and pressure. (The sandstone originally was produced by the compression of sand and mud layers.)

The Sykesville Formation is a metagraywacke "matrix" that contains chunks of other rocks, including gneiss, granite, carbonate, schist, and amphibolite. Metagraywacke is a dark gray metamorphic rock that has long been quarried along the Potomac River. First mined in 1850, this rock is probably what was once called "Potomac bluestone." Sykesville gneiss was once the local stone to use for foundations and rubble and was the primary material used to build many famous buildings. "Potomac bluestone" was used to build Georgetown University's Healey Building, St. Elizabeths Hospital, the seawall at Hains Point (East Potomac Park), the old stone gatehouse at Constitution Avenue and 17th Street, N.W., as well as a number of prominent structures at the Smithsonian's National Zoological Park, including the Panda House, the original Elephant House, and the Mane Restaurant. It was also used in the construction of many houses in the city's northwest quadrant, including the Old Stone House, the city's oldest standing building,

Rocks of the Sykesville Formation, near the C&O Canal and along the shores of the Potomac River above the Fall Line.

Laurel Formation rocks near Boulder Bridge, in Rock Creek Park.

which is now administered by the National Park Service. The house, located in Georgetown at 3051 M Street, N.W., was built in 1765 of local "blue granite."

Common Locations   You can see Sykesville Formation rocks on the north end of Roosevelt Island and just north of Chain Bridge, between the Potomac River and the C&O Canal, where acres of outcrop are available for exploration.

KEY POINTS

- A similar formation, the Laurel Formation, was named for the town of Laurel, Maryland. Many of the rock outcroppings you see near Boulder Bridge in Rock Creek Park belong to this formation. Like the Sykesville Formation, these rocks are ancient oceanic sediments that were "cooked" during the Taconian Orogeny.

Georgetown Intrusive Suite

ETYMOLOGY

Named for Georgetown. It is a "suite," or group of rocks formed by the intrusion of magma into metamorphic "host" rocks there.

Natural History   During the Taconian Orogeny, some sediments not only metamorphosed, they partially melted, producing magma. This liquid rock forced its way into other rocks through fractures, in a process called intrusion. This is how igneous rocks in this area were formed.

Description   The dark rock shown here is gabbro, an igneous rock formed by slow cooling of underground molten rock, or magma. To geologists, the darkness indicates that this rock has elevated levels of iron and magnesium. (The Vietnam Veterans Memorial is also made of gabbro.)

In this photo, you see a vein of milky quartz. How did this white vein come to "decorate" the dark gabbro? One look tells a geologist very different times and circumstances brought these two together.

After the gabbro had formed, it fractured. At some point, hot water carried dissolved silica into the cracks. As the temperature cooled, the silica precipitated or crystallized out of the water, and the cracks were basically sealed shut by a crystalline glue. The milky color shows that the crystallization process was not perfect—the white color indicates that many tiny water bubbles were left. These tiny imperfections scatter incoming light, producing the white color.

In other outcrops in this area, geologists find lighter-colored granites intruded into the dark gabbro.

Common Locations   You can see evidence of the Georgetown Intrusive Suite along Canal Road, N.W., from its intersection with Foxhall Road, N.W., east to Key Bridge. Also, outcrops can be found in Rock Creek Park, in a section just north of Montrose Park.

KEY POINTS

- Georgetown is not the only area where intrusive magma played a major role in rock formation. Kensington tonalite (a light-colored igneous rock found in Rock Creek Park) and the Dalecarlia Intrusive Suite (in the area of Dalecarlia Reservoir in the city's far northwest corner), provide other important examples of igneous intrusions.

Rocks of the Georgetown Intrusive Suite, near Canal Road and the C&O Canal.

Coastal Plain Deposits

Natural History   It is hard to find exposed evidence of ancient Coastal Plain deposits, but if you look at the exposed flats along the Anacostia and Potomac shorelines below the Fall Line, you get some idea of how these deposits were made.

Description   Clay and silt dominate the river deposits along the city's river shores. These sediments are products of the breakdown of exposed rock upstream, but from where exactly? That's a good question. If you bend down and scoop up some Potomac shore mud in your hand, particles you hold might be from a few yards upstream or as far away at the river's Appalachian headwaters.

Common Locations   Check out the shoreline and marsh along Roosevelt Island's south side, around the Tidal Lock near the Watergate, or along the Anacostia River shoreline.

KEY POINTS

- During the mid-Cretaceous period the area we now call Washington, D.C., was ocean shore.
- Coastal Plain gravel and other sediments were laid down much more recently than the exposed Piedmont rocks you see above the Fall Line. Most of it consists of river and ocean deposits laid down from as far back as 100 million years ago to as recently as about 10,000 years ago.

Other Major Geological Events

The Taconian Orogeny was a major geologic event, but two other mountain-building events helped create the Appalachians and shape the city's landscape. (Today, the Appalachians are but a vestige of their former grandeur.) First, the Acadian Orogeny took place around 360 million years ago, due to the collision of eastern North America with a microcontinent called Avalonia. Then, between around 300 million and 250 million years ago, the Alleghanian Orogeny occurred when North America collided with Africa. This colossal event resulted in the formation of the huge landmass Pangaea. When this event was done, the Appalachians were at their highest, a range as dramatic as the Himalayas, with its tallest peaks running through where the city stands today. As soon as the tectonic forces waned, the mountains stopped growing and began to wear away.

Between 200 and 180 million years ago, the huge landmass Pangaea began to break apart, in the process ripping apart the mountain

Coastal Plain soils exposed at low tide along the south end of Roosevelt Island, just below the Fall Line. The constant movement of sediments down the Potomac River here hints at the millions of years of oceanic and river deposits that built the Coastal Plain.

range and slowly sending the other half drifting east, where it sits today along Africa's northwest coast.

By about 100 million years ago, during the middle of the Cretaceous period, the mountains had worn down. Ocean shore sat where the city is now and meandering rivers began depositing sediments in layers atop the eroded mountains' deep roots. Around Mount Pleasant and some other areas in the city, Cretaceous shoreline gravel deposits are exposed. They are distinctly light orange, with well-rounded white quartzite cobbles. These were the source of some of the cobbles used to pave Georgetown streets.

Some time later, rivers began to cut down again. This action removed the upper sedimentary layers and cut deep into the igneous and metamorphic rocks hidden beneath. As a result, the landscape rose into high relief once more, creating dra-matic landscapes such as the Potomac River Gorge and Rock Creek Valley.

ORGANIZATIONS IN AND AROUND WASHINGTON, D.C.

Anacostia Watershed Society: http://www.anacostiaws.org

Audubon Naturalist Society: http://www.audubonnaturalist.org

Casey Trees: http://www.caseytrees.org/

Cherry Blossom Information: http://www.nps.gov/cherry/cherry-blossom-bloom.htm (National Park Service bloom schedule for cherry blossoms)

http://www.nps.gov/cherry (Cherry Blossom Festival); http://www.nps.gov/cherry/cherry-blossom-maps.htm (maps)

http://www.nationalcherryblossomfestival.org/ (National Cherry Blossom Festival)

City Wildlife: http://www.citywildlife.org/

Coalition for the Capital Crescent Trail: http://www.cctrail.org/

District Department of the Environment: http://green.dc.gov/service/fisheries-and-wildlife

District Department of Transportation Urban Forestry Program: http://ddot.dc.gov/DC/DDOT/On+Your+Street/Urban+Forestry

The Entomological Society of Washington: http://entsocwash.org

Friends of the National Arboretum: http://www.fona.org

Maryland & D.C. Birding: http://www.mdbirding.com/

Maryland Native Plant Society invasive plants projects (includes Washington, DC): http://mdflora.org/invasiveprojs.html

Maryland Ornithological Society: http://www.mdbirds.org

National Capital Planning Commission: http://www.ncpc.gov/

Potomac Conservancy: http://www.potomac.org/site/

Rock Creek Conservancy: http://www.rockcreekconservancy.org

Rock Creek Park: http://www.nps.gov/rocr/

Sierra Club, Washington, D.C., Chapter: http://www.dc.sierraclub.org/

Smithsonian National Museum of Natural History: http://www.mnh.si.edu/

Smithsonian National Zoological Park and Friends of the National Zoo: http://nationalzoo.si.edu/

U.S. National Arboretum: http://www.usna.usda.gov/

http://www.ars.usda.gov/Aboutus/docs.htm?docid=4745 (frequently asked questions)

Virginia Herpetological Society: http://www.virginiaherpetologicalsociety.com/

Virginia Society of Ornithology: http://virginiabirds.net/

TRANSPORTATION RESOURCES

Anacostia Riverwalk Trail: http://www.anacostiawaterfront.org/awi-documents/anacostia-riverwalk-trail-documents/dc-anacostia-riverwalk-trail-map/

Capital Crescent Trail: http://www.cctrail.org

DC Circulator Bus Routes: http://www.dccirculator.com/Home/BusRoutesandSchedules.aspx

Metrobus and Metrorail: MetroOpensDoors.com

BIBLIOGRAPHY

Berman, Richard L., and Deborah McBride. *Natural Washington.* Charlottesville, VA: EPM, 1999.

Blond, Becca, and Aaron Anderson. *Washington, DC: City Guide.* Oakland, CA: Lonely Planet, 2007.

Brauning, Daniel W. *Atlas of Breeding Birds in Pennsylvania.* Pittsburgh, PA: University of Pittsburgh Press, 1992.

Brinkley, Edward S. *National Wildlife Federation Field Guide to Birds of North America.* New York: Sterling, 2007.

Brock, Jim P., and Kenn Kaufman. *Butterflies of North America.* New York: Houghton Mifflin, 2003.

Buchsbaum, Ralph, Mildred Buchsbaum, John Pearse, and Vicki Pearse. *Animals without Backbones.* Chicago: University of Chicago Press, 1987.

Choate, Ernest A. *The Dictionary of American Bird Names.* Boston: Gambit, 1973.

Choukas-Bradley, Melanie. *An Illustrated Guide to Eastern Woodland Wildflowers and Trees: 350 Plants Observed at Sugarloaf Mountain, Maryland.* Charlottesville, VA: University of Virginia Press, 2004.

Choukas-Bradley, Melanie. *City of Trees.* Charlottesville, VA: University of Virginia Press, 2008.

Choukas-Bradley, Melanie. "Historic Trees & Gardens of 1600 Pennsylvania Avenue." *American Forests,* Summer 2010.

Conant, Roger, and Joseph T. Collins. *A Field Guide to Reptiles and Amphibians: Eastern and Central North America.* New York: Houghton Mifflin, 1998.

Conniff, Richard. "Swamp Thing: Unmasking the Snapping Turtle." *National Geographic,* March 1999.

Day, Leslie. *Field Guide to the Natural World of New York City.* Baltimore: Johns Hopkins University Press, 2007.

Dunkel, Sidney W. *Dragonflies through Binoculars.* New York: Oxford University Press, 2000.

Dunn, Jon L., and Jonathan Alderfer. *National Geographic Field Guide to the Birds of North America.* Washington, D.C.: National Geographic, 2011.

Dunn, Jon L., and Kimball L. Garrett. *A Field Guide to Warblers of North America.* New York: Houghton Mifflin, 1997.

Elias, Thomas S. *The Complete Trees of North America.* New York: Gramercy, 1987.

Ellison, Walter G. *Second Atlas of the Breeding Birds of Maryland and the District of Columbia.* Baltimore: Johns Hopkins University Press, 2010.

Elphick, Chris, John B. Dunning Jr., and David A. Sibley. *The Sibley Guide to Bird Life & Behavior*. New York: Alfred A. Knopf, 2001.

Ernst, Ruth S. *The Naturalist's Garden*. Old Saybrook, CT: Globe Pequot Press, 1993.

Evans, Arthur V. *Field Guide to Insects and Spiders of North America*. New York: Sterling, 2007.

Fergus, Charles. *Wildlife of Virginia and Maryland and Washington, D.C.* Mechanicsburg, PA: Stackpole Books, 2003.

Fisher, Alan. *Country Walks Near Washington*. Baltimore: Rambler Books, 1996.

Fleming, Cristol, Marion B. Lobstein, and Barbara Tufty. *Finding Wildflowers in the Washington-Baltimore Area*. Baltimore: Johns Hopkins University Press, 1995.

Garber, Steven D. *The Urban Naturalist*. Mineola, NY: Dover, 1987.

Garland, Mark S. *Watching Nature: A Mid-Atlantic Natural History*. Washington, D.C.: Smithsonian Institution Press, 1997.

Gilbert, Carter R., and James D. Williams. *National Audubon Society Field Guide to Fishes*. New York: Alfred A. Knopf, 2002.

Glickman, Barbara, and Valerie Brown. *Capital Splendor: Gardens and Parks of Washington, D.C.* Woodstock, VT: Countryman Press, 2012.

Guglielmino, Janine. "Natural Capital: Trees of Washington, D.C." *American Forests*, Spring 2001.

Halle, Louis J. *Spring in Washington*. Baltimore: Johns Hopkins University Press, 1957.

Halle, Louis J. "The Veery Breeding in Washington, D.C." *Auk*, vol. 60, January 1943.

Harrison, Colin. *A Field Guide to the Nests, Eggs and Nestlings of North American Birds*. Cleveland, OH: Collins, 1978.

High, Mike. *The C&O Canal Companion*. Baltimore: Johns Hopkins University Press, 2000.

Holloway, Joel E. *Dictionary of Birds of the United States*. Portland, OR: Timber Press, 2003.

Iliff, Marshall J., Robert F. Ringler, and James L. Stasz. *Field List of the Birds of Maryland*. Baltimore: Maryland Ornithological Society, 1996.

Kaufman, Kenn. *Lives of North American Birds*. New York: Houghton Mifflin, 1996.

Kays, Roland W., and Don E. Wilson. *Mammals of North America*. Princeton, NJ: Princeton University Press, 2002.

Martin, Edwin M. *A Beginner's Guide to Wildflowers of the C&O Towpath*. Washington, D.C.: Smithsonian Institution Press and the Audubon Naturalist Society of the Central Atlantic States, 1984.

Means, John. *Roadside Geology of Maryland, Delaware, and Washington, D.C.* Missoula, MT: Mountain Press, 2010.

Mitchell, Joseph C. *The Reptiles of Virginia.* Washington, D.C.: Smithsonian Institution Press, 1994.

National Park Service. *Cultural Landscapes Inventory: D.C. War Memorial, National Mall & Memorial Parks.* Washington, D.C.: National Park Service, 2009.

NBS/NPS Vegetation Mapping Program. *Vegetation Classification of Rock Creek Park.* Boston: The Nature Conservancy, 1998.

Neal, Bill. *Gardener's Latin.* Chapel Hill, NC: Algonquin Books of Chapel Hill, 1992.

Oman, Anne H. *Twenty-Five Bicycle Tours in and around Washington, D.C.* Woodstock, VT: Backcountry.

Page, Lawrence M., and Brooks M. Burr. *A Field Guide to Freshwater Fishes: North America North of Mexico.* Boston: Houghton Mifflin, 1991.

Peterson, Lee. *A Field Guide to Edible Wild Plants of Eastern and Central North America.* Boston: Houghton Mifflin, 1977.

Peterson, Roger T. *A Field Guide to the Birds East of the Rockies.* Boston: Houghton Mifflin, 1980.

Plant Names Explained. Boston: Horticulture Publications, 2005.

Rappole, John H. *Birds of the Mid-Atlantic Region and Where to Find Them.* Baltimore: Johns Hopkins University Press, 2002.

Robbins, Chandler S., and Eirik A. T. Blom. *Atlas of the Breeding Birds of Maryland and the District of Columbia.* Pittsburgh, PA: University of Pittsburgh Press, 1996.

Roberson, Mary-Russell, "Beneath It All: The Geology of the Zoo," adapted from an article that appeared in *ZooGoer*, 1988.

Russell, Bill. *Field Guide to Wild Mushrooms of Pennsylvania and the Mid-Atlantic.* University Park: Pennsylvania State University Press, 2006.

Schmidt, Martin F., Jr. *Maryland's Geology.* Centreville, MD: Tidewater, 1997.

Shelton, Napier, "A Walk through Washington Woods and ANS History." *Audubon Naturalist News*, February/March 2008.

Sibley, David A. *The Sibley Field Guide to Birds of Eastern North America.* New York: Alfred A. Knopf, 2003.

Sibley, David A. *The Sibley Guide to Trees.* New York: Alfred A. Knopf, 2009.

Smith, A. W. *A Gardener's Handbook of Plant Names: Their Meanings and Origins.* Mineola, NY: Dover, 1997.

Spilsbury, Gail. *Rock Creek Park.* Baltimore: Johns Hopkins University Press, 2003.

Swearingen, Jil M., Kathryn Reshetiloff, Britt Slattery, and Susan M. Zwicker. *Plant Invaders of Mid-Atlantic Natural Areas.* Washington, D.C.: National Park Service and U.S. Fish & Wildlife Service, 2002.

Swearingen, Jil M., and K. Saltonstall. Phragmites *Field Guide: Distinguishing Native and Exotic Forms of Common Reed* (Phragmites australis) in the United States. Plant Conservation Alliance's Weeds Gone Wild Website, http://www.nps.gov/plants/alien/pubs/index.htm.

Wheeler, Linda. "Beaver Continues to Dine on Tidal Basin." *Washington Post*, April 8, 1999, A1, http://www.washingtonpost.com/wp-srv/local/daily/april99/beaver8.htm.

Wherry, Edgar T. *The Fern Guide: Northeastern and Midland United States and Adjacent Canada.* Mineola, NY: Dover, 1995.

Whitaker, John O., Jr. *National Audubon Society Field Guide to North American Mammals.* New York: Alfred A. Knopf, 1997.

Wilds, Claudia. *Finding Birds in the National Capital Area.* Washington, D.C.: Smithsonian Institution Press, 1992.

Williams, Michael D. *Identifying Trees: An All-Season Guide.* Mechanicsburg, PA: Stackpole Books, 2007.

WEBSITES

http://bna.birds.cornell.edu/bna (The Birds of North America Online)

http://www.marylandinsects.com/index.html (Mid-Atlantic Invertebrate Field Studies)

http://www.mdbirds.org/sites/dcsites/dcbirds.html (Maryland Ornithological Society's online guide to birding sites in Washington, D.C.)

http://nationalzoo.si.edu/AboutUs/History/beneathitall.cfm (geology of the National Zoo)

http://www.nps.gov/nama/ (National Mall and Memorial Parks)

http://www.nmca.org/PHRAG_FIELD_GUIDE.pdf (*Phragmites* field guide)

http://www.nps.gov/plants/alien (Weeds Gone Wild)

http://nas.er.usgs.gov/queries/factsheet.aspx?speciesID=4 (common carp fact sheet)

http://www.lostladybug.org/file_uploads/FieldGuide.pdf (Lost Ladybug Project identification guide)

http://www.dnr.state.md.us/fisheries/fishfacts/index.asp (Maryland Department of Natural Resources, Fish Facts)

http://www.montgomerycountymd.gov/dectmpl.asp?url=/content/dep/water/monBioCrayfish.asp (Montgomery County, Maryland, crayfish overview)

http://www.thenaturalcapital.com/ (The Natural Capital: Getting Outside,
    Inside the Beltway, a website authored by Elizabeth Hargrave and Matt
    Cohen)

http://vulcan.wr.usgs.gov/LivingWith/VolcanicPast/Places/volcanic_
    past_washington_DC.html (America's Volcanic Past: Washington,
    D.C.)

http://vulcan.wr.usgs.gov/LivingWith/VolcanicPast/Places/volcanic_
    past_maryland.html ("America's Volcanic Past: Maryland.")

http://www.nvcc.edu/home/cbentley/dc_rocks/ (Callan Bentley's D.C.
    Rocks, Northern Virginia Community College)

http://gloverparkhistory.com/estates-and-farms/alliance-farm/
    charles-carrol-glover/

http://gloverparkhistory.com/estates-and-farms/hillandale/anne-
    archbold/ (Historical Sketches of Glover Park, Upper Georgetown,
    and Georgetown Heights, by Carlton Fletcher)

http://intheplayingfields.tumblr.com/post/1295794855/the-past-
    lies-sleeping-seeking-the-mason-house ("After Dark in the Playing
    Fields," a search for Mason's mansion on Roosevelt Island)

http://www.loc.gov/exhibits/treasures/tri001.html (Library of Congress:
    Pierre-Charles L'Enfant's city plan)

INDEX

The letter "f" following a page number indicates a photo, "pl" indicates a plate. and "m" indicates a map.

INDEX

381

HOWARD YOUTH writes about wildlife and conservation issues for a variety of publications, including *Bird Watcher's Digest*, where he is a field editor. He holds a journalism degree from the University of Maryland and worked as an editor at Worldwatch Institute and Friends of the National Zoo. An avid naturalist since age nine, he grew up studying the animals and plants he saw in Washington, D.C., and its suburbs. Although he has traveled the world, writing about conservation issues in India, Spain, Honduras, Malta, and Ecuador, he is most at home exploring the parks in and around the nation's capital.

MARK A. KLINGLER is a scientific illustrator at Carnegie Museum of Natural History in Pittsburgh, Pennsylvania. He studied at Carnegie Mellon University and Pennsylvania Academy of the Arts. His award-winning work has appeared in books, major scientific journals, and popular magazines and has been featured in numerous museums and galleries. He enjoys teaching and presenting workshops about wildlife and conservation art.

ROBERT E. MUMFORD, JR., is a freelance nature photographer. His work has appeared in many of the best-known American bird and nature magazines. He is the author of the coffee-table book, *Spring Comes to Washington*. He lives in the woods 25 miles from the White House, with mostly deer, squirrels, raccoons, and birds as neighbors.

GEMMA RADKO who designed the maps for this book, is a graphic designer at American Bird Conservancy, where she designs and helps write member publications. She loves to watch birds in Washington, D.C., and everywhere else.

About the type The text of this book is set in Andrade, designed by Dino dos Santos in 2005. This typeface is inspired by the typographic work of Manoel de Andrade de Figueiredo (1670–1735). The work of Andrade de Figueiredo is among the most impressive examples of type design from the eighteenth century. Updated to reflect contemporary sensibilities, Andrade is a tribute to Portuguese typography.